Micro-grids - Applications, Operation, Control and Protection

Edited by Mahmoud Ghofrani

Published in London, United Kingdom

IntechOpen

Micro-grids - Applications, Operation, Control and Protection
http://dx.doi.org/10.5772/intechopen.77550
Edited by Mahmoud Ghofrani

Contributors
Gaspard D\'Hoop, Olivier Deblecker, Leila Ghomri, Mounir Khiat, Sid Ahmed Khiat, Mohamed Mankour, Edward Donahue, Ali Hadi Abdulwahid, Andreea El-Leathey, Ramamoorty M, S.V.N.L. Lalitha

First published in London, United Kingdom, 2019 by IntechOpen
IntechOpen is the global imprint of INTECHOPEN LIMITED, registered in England and Wales, registration number: 11086078, The Shard, 25th floor, 32 London Bridge Street
London, SE19SG - United Kingdom
Printed in Croatia

British Library Cataloguing-in-Publication Data
A catalogue record for this book is available from the British Library

Additional hard and PDF copies can be obtained from orders@intechopen.com

Micro-grids - Applications, Operation, Control and Protection
Edited by Mahmoud Ghofrani
p. cm.
Print ISBN 978-1-78984-061-2
Online ISBN 978-1-78984-062-9
eBook (PDF) ISBN 978-1-78985-442-8

For EU product safety concerns:
IN TECH d.o.o., Prolaz Marije Krucifikse Kozulić 3, 51000 Rijeka, Croatia, info@intechopen.com or visit our website at intechopen.com.

Supporting open minds since 2005

Meet the editor

Mahmoud Ghofrani received his BSc degree from Amirkabir University of Technology and his MSc degree from the University of Tehran, both degrees in Electrical Power Engineering. He received his PhD degree from the University of Nevada, Reno, in 2014. Dr. Ghofrani is currently an assistant professor at the School of Science, Technology, Engineering and Mathematics at the University of Washington, Bothell. His research interests include power systems operation and planning, renewable energy systems, smart and micro-grids, electric vehicles, and electricity markets. He is the author and coauthor of more than 50 publications in top-tier journals and conferences. In addition, he has five book chapter publications.

Contents

Preface

The integration of recent and emerging energy technologies in the existing electric grid requires modifications in several aspects of the grid, including its architecture, protection, operation, and control. Micro-grid provides a solution for integrating distributed energy resources (DERs) such as renewable energy generation, energy storage systems, electric vehicles, controllable loads, etc. and delivers flexibility, security, and reliability by operating in both grid-connected and isolated modes.

This book provides an overview of micro-grid solutions, applications, and implementations. State-of-the-art methods for micro-grid operation, optimization, and control are presented. Distributed energy resources and their interactions in micro-grids are also studied. In addition, micro-grid designs and architectures are covered, as are micro-grid protection strategies and schemes for different operation modes. This book presents a new model for urban development using micro-grid features and technologies. Furthermore, IEEE Standards for AC and DC micro-grids are investigated.

The book is organized in three sections and six chapters.

Section 1 includes three chapters covering micro-grid implementations for different applications, while Section 2 includes a chapter covering optimization and control of micro-grids for power quality issues.

Finally, Section 3 includes two chapters covering protection of micro-grids with DERs.

Chapter 1 develops a novel model for urban development and design based on micro-grid solutions. Several technologies such as conventional fossil-fueled and steam-powered generation, renewable wind and solar systems, and energy storage devices are included in the model with the objective of optimizing the building design to maximize such assets. A commercial tower with real-world data is studied to demonstrate the economic and environmental advantages of the proposed development model over traditional developments.

The model is based on trigeneration of power. Conventional generators produce electricity and heat at the first stage, which is used to power a steam turbine to produce more electricity at the second stage. The third stage of trigeneration is to use the remaining exhaust heat to provide building heat, hot water, and air conditioning. Renewable solar and wind power supplement the system. The design includes energy storage systems to help balance the power over peak and off-peak load periods.

The study demonstrates that the proposed model balances load and generation capability on a large scale, allowing the construction of large numbers of buildings to accommodate increasing populations with essentially no impact on the existing power distribution infrastructure. It is also environmentally sustainable, producing fewer emissions than a traditional development on the same scale. The economic and environmental advantages of this development model provide an opportunity to promote general welfare.

Chapter 2 presents a model for the development of micro-grids in remote areas. The model includes diesel generators and renewable wind and solar. A real-time simulation is performed to evaluate different scenarios for micro-grid operation, and to determine the best scenario enabling the micro-grid to operate in an island mode without the need to connect to the main grid.

Lack of commercially available electronic devices and systems, and their required standards and regulations, is a major challenge that hinders the rapid development of residential DC micro-grids. Chapter 3 studies the existing standards applicable to DC micro-grids and establishes the main technical requirements and characteristics for the implementation of DC systems designed for residential applications. Advantages and challenges of low-voltage DC distribution systems, particularly when integrating renewable energy sources along with storage systems, are discussed in the chapter. In addition, several topologies are presented for residential DC micro-grids to address the challenges.

Chapter 4 evaluates micro-grid power quality under steady-state grid-connected operation. An energy management method is proposed to optimize generation, storage, and load scheduling to resolve the power quality issues of the micro-grid. A regulation loop between a mixed-integer linear programming optimization and harmonic load flow is used to adjust the power generation, storage, and consumption in response to the electricity price of the grid while meeting the power quality standards. A test micro-grid is used to evaluate the proposed method for different scenarios. The results demonstrate the efficiency of the designed energy management.

Increased penetration of DERs has complicated the protection of micro-grids. The conventional protection schemes are mainly designed for radial distribution systems and pose serious challenges when applied to mesh-connected distribution systems with DERs.

Chapter 5 provides a comprehensive analysis of micro-grid protection systems. Several methods, including directional over current relays, distance relays, differential protection, voltage-based protection, transformation method, harmonic method, and adaptive protection schemes, are presented. These methods are evaluated for their effective implementation with minimal changes in the existing infrastructure.

Chapter 6 proposes a novel differential protection scheme based on current sensors for micro-grids. The proposed method is modeled in Simulink/Matlab and evaluated for several different faults, including single line-to-ground, line-to-line, double line-to-ground, three-line, and three line-to-ground faults. Simulation results demonstrate the efficiency of the proposed differential protection over distance relay protection due to its reduced number of input data and computational complexity. It is concluded that differential protection is the preferred solution for micro-grid applications.

Dr. Mahmoud Ghofrani
University of Washington Bothell, USA

Section 1

Applications and Implementations

Chapter 1

Microgrids: Applications, Solutions, Case Studies, and Demonstrations

Edward J. Donahue

Abstract

Rapid urbanization of the world's population is creating great sociological, environmental, and structural strains on the cities where people are moving to. Housing is becoming scarce and expensive, while the need to build new housing is placing great burdens on existing infrastructure—especially local power grids. It will be shown that integrating urban development around a microgrid concept would greatly alleviate the problems associated with urbanization. Incorporation of a microgrid, based on a cogenerating power station where waste heat is used to provide climate control and hot water and where power production is supplemented with renewable energy sources, would effectively remove the development from the local grid and greatly reduce greenhouse gas emissions. Additionally, this model can accommodate any combination of large-scale residential, commercial, or industrial developments to revitalize the local neighborhood and can do so at a level of profit that would allow for lower rents, creating housing and job opportunities for those who are most in need.

Keywords: microgrid, urban development, cogeneration, trigeneration, renewable energy

1. Introduction

Urbanization has occurred throughout history as agricultural societies evolve [1, 2]. The concentration of population (Pop.) leads to a specialization of labor, allowing individuals to concentrate their efforts into fields where they have a particular aptitude. This inevitably leads to the rise of some type of market economy in which one trades upon the skills possessed to fulfill needs in areas outside of one's chosen field of endeavor. Urbanization historically has led to greater overall prosperity in the long term [3–6]. However, immediate consequences are more varied and lead to the "known evils" of city life: poverty, slums, an uneven distribution of resources, and a marked decline in public health [7].

The historical trend toward urbanization is continuing and accelerating into the present. The world population has grown dramatically in the past 75 years and has become increasingly urbanized. The total population of the planet grew by 148% between 1960 and 2017 and by 42% in the roughly quarter century between 1990 and 2017 [8]. During that same quarter century period, the urban population of the planet grew at almost double the rate of the overall population, increasing by 83%

between 1990 and 2017 [9]. In 1990, 43% of the world's population lived in urban centers compared to 54% of a larger population in 2017, an increase of 1.9 billion people occupying the world's cities [8, 9].

The link between urbanization and the decline of public health has been weakened by advances in basic sanitation and the developments in modern medicine. There is now no discernable difference in life expectancy and infant mortality between urban and rural areas in developed countries, and metrics now favor the urban population in many developing nations [4, 6, 10]. However, the unequal distribution of resources still persists in urban centers, especially with regard to inequalities in the cost and quality of housing. The modern age has added energy to the list of resources whose availability is uneven and prosperity related [11–13]. This chapter will present a model for alleviating these systemic inequalities through the incorporation of electric microgrids directly into the planning and construction of new urban developments.

The United States Department of Energy defines a microgrid as "A group of interconnected loads and distributed energy resources that act as a single controllable entity with respect to the grid. A microgrid can connect and disconnect from the grid to enable it to operate in both a grid-connected or island mode" [14]. A model is developed wherein a trigenerating, combined cycle electrical generating system is integrated into the design and construction of a combined residential (Res.) and commercial (Com.) development project. The term combined cycle indicates that steam produced as exhaust from a fossil fuel-powered turbine operates an additional steam turbine in order to increase efficiency. It is referred to as "trigenerating" because the waste heat from the combined cycle is then used to provide heat, hot water, and air conditioning (AC) to buildings on the microgrid, further increasing efficiency. The model also incorporates renewable energy sources, solar panels and wind turbines, in the building structures.

It will be shown that an integrated development is economically and environmentally sustainable and is also profitable. The integrated development will be modeled in several cities around the world which were selected in order to present a representative cross section of both environmental conditions and levels of national economic development. In developed countries, the implementation of the methodology presented will alleviate the strain on the now-aging electrical grids that accelerated urban development is causing. In less developed countries, its adoption will add to often inadequate supply. Local conditions of cost, revenue, and environment are incorporated into each model.

2. Cities selected

The following cities were selected for inclusion in this study: Cairo, Egypt; Lagos, Nigeria; Shanghai, China; Mumbai, India; London, England; New York City (NYC), United States; and Mexico City, Mexico. These cities were chosen for the following reasons. They all are considered "megacities" as defined by the United Nations, with populations greater than 10 million [15]. As seen in **Table 1**, they have all had major population increase over the past 20 years [16]. **Table 1** also shows their ranking by population globally (WPR) and with respect to their respective continents (CPR; North America, NA; South America, SA) [17, 18]. Additionally, **Table 1** shows that they are located in varied Köppen-Geiger (K-G) climate zones, a fact that affects heating and air-conditioning loads and cycles [19]. The definition of the K-G climate zones is given in **Table 2** [20].

The cities vary greatly in their wealth and development. This impacts the reliability of the electrical supply and the availability of affordable housing. **Table 3**

City	Continent	Pop.	Pop.	Change	WPR	CPR	K-G class
Cairo	Africa	9,900,000	18,800,000	89.9%	8	1	Bwh
Lagos	Africa	4,800,000	12,200,000	154.2%	21	2	Aw
Shanghai	Asia	8,600,000	23,500,000	173.3%	3	1	Cfa
Mumbai	Asia	12,400,000	19,300,000	55.7%	6	3	Aw
London	Europe	6,800,000	8,700,000	28.0%	38	3	Cfb
Mexico City	NA	15,600,000	21,300,000	36.5%	4	1	Cwb
NYC	NA	16,100,000	18,600,000	15.5%	9	2	Cfa
Sao Paolo	SA	14,800,000	20,900,000	41.2%	5	2	Cfa

Table 1.
Population and climate of cities of interest.

Main climates		Precipitation		Temperature			
A	Equatorial	W	Desert	h	Hot arid	c	Cool summer
B	Arid	S	Steppe	k	Cold arid	d	Extremely continental
C	Warm	f	Fully humid	a	Hot summer	F	Polar frost
D	Snow	s	Summer dry	b	Warm summer	T	Polar tundra
E	Polar	w	Winter dry				

Table 2.
Köppen-Geiger climate classification.

City	Country	HDI Index	NP (%)	LCS (%)
Cairo	Egypt	0.696	25%	10.6%
Lagos	Nigeria	0.532	46%	66%
Shanghai	China	0.752	4.6%	N/A
Mumbai	India	0.640	22%	41.3%
London	United Kingdom	0.922	N/A	27%
Mexico City	Mexico	0.774	52.3%	40%
New York City	United States	0.924	N/A	20%
Sao Paolo	Brazil	0.759	8.9%	19%

Table 3.
State of economic development for cities of interest.

presents the state of economic development in the nations in which these cities are located, as measured by the United Nations Human Development Index (HDI) [21] as well as the percentage of the national population living in poverty (NP%) [22]. The percentage of any given city's population living in slums is not presented in a self-consistent manner. The United Nations defines a slum household as "a group of individuals living under the same roof in an urban area who lack one or more of the following: Durable housing of a permanent nature that protects against extreme climate conditions; sufficient living space which means not more than three people sharing the same room; easy access to safe water in sufficient amounts at an affordable price; access to adequate sanitation in the form of a private or public toilet shared by a reasonable number of people and security of tenure that prevents forced evictions" [23]. The world organizations do not keep data on such a granular level, and national data might not report poverty in terms of locality. As the purpose of this study is to use sustainable development to improve living conditions, the state of local housing quality is of prime interest. Therefore, it was deemed appropriate to use non-internally consistent data for local slum conditions in **Table 3** with data on the percentage of the population living in slum conditions for each city (LSC%) which was obtained from the following sources: Cairo [24], Lagos [25], Mumbai [26], London [27], Mexico City [28], New York City [29], and Sao Paolo [30]. There is no measure or recognition of slum conditions in Shanghai.

The data on electrical distribution and reliability shown in **Table 4** correlates strongly with the economic prosperity of the country wherein that city is located, as well as the age of the supporting infrastructure. The National Access to Electricity for 2016 (NAE) [31] and the National Average Blackout Days per Month (BD/M) [32] are strong indicators of development. Both the National Quality of Electricity Supply [33] and the National Average Interruption Frequency Index [34] are reported using the Reliability of Supply and Transparency of Tariff Index, a scale which "encompasses quantitative data on the duration and frequency of power outages as well as qualitative information on how utilities and regulators handle power outages and how tariffs and tariff changes are communicated to customers" [35]. A score of 8 is the highest possible on this scale. The measurement of power transmission and distribution losses (PD/T) is presented as an indicator of the existing strain on the local distribution networks [36].

A comparison of **Tables 3** and **4** shows a strong correlation between the National HDI Index and the quality of electricity distribution as measured by both the Quality of Electric Supply Index and the Average Interruption Frequency Index. The state of the electricity distribution grids servicing the cities cited in this work fit into three categories: insufficiently maintained and planned (Cairo [37] and Lagos [38]), extensive but aging (London [39], Mexico City [40], New York City [41], and Sao Paolo [42]), and relatively new and robust (Mumbai [43] and Shanghai [44]). The categorization broadly mirrors HDI in the nations in which the selected cities are located. China and India are rapidly modernizing from an underdeveloped base and can build or expand a modern, robust grid from scratch. The United States and United Kingdom, and, to a lesser extent, Mexico and Brazil, have long established industrial economies, meaning that increasing rate of urbanization is a straining and extensive, but aging, infrastructure. Egypt and Nigeria are underdeveloped countries relying on insufficient base infrastructure.

City	Country	NAE	BD/M	QES	AIF	PD/T
Cairo	Egypt	100%	1.8	5	3	14%
Lagos	Nigeria	59.3%	32.8	1.8	0	16%
Shanghai	China	100%	0.1	3.9	6	5%
Mumbai	India	84.5%	13.8	3.1	7	19%
London	United Kingdom	100%	0	6.7	6	12%
Mexico City	Mexico	100%	1.6	4.1	7	14%
NYC	United States	100%	0	6.3	7.2	6%
Sao Paolo	Brazil	100%	1.6	5	5.4	16%

Table 4.
Quality of electrical supply at the national level.

3. Microgrids

Growing metropolitan areas require greater local power generation capacity in order to meet growing local needs and to maintain balance in the national distribution grids. However, the fact that this energy is needed in already congested cities presents an economic problem. Reliable energy is necessary for sustained growth, but the real estate needed for additional power production facilities is also needed for further housing and commercial uses. The use of land for power production addresses a potentially catastrophic future problem, while development for residential and commercial use produces profits for developers and increased tax bases for the municipality. Barring direct government intervention, the latter is the predominantly preferred course of action.

Both needs can be simultaneously addressed through integrated development. The following sections will outline how such a development might be structured as well as the economic and ecological return produced. Although the definition of a microgrid [14] seems straightforward, this definition relies largely on self-classification and makes actual quantification difficult. The data available at mic rogridprojects.com, a trade-related site that is partially based upon self-reporting, illustrates the elasticity of the definition [45]. A majority of microgrids are located in remote, undeveloped areas or on distant islands, places where connecting to the distribution grid is economically unviable or even physically impossible, making local generation the only possible choice. A prime example of this is the fact that 816 MW of the total 844 MW generated in remote areas of Asia is generated by the Russia Far-East Microgrid Portfolio, a conglomeration of 82 generating station serving remote and isolated communities in Siberia which could, in fact, be considered a proper power distribution grid in its own right. Also, the municipal adoptions of microgrids in North America are illustrative of the inherent idiosyncrasies. Of 114.3 total MW generated in this sector, 104 are generated by the New Jersey Transit microgrid. The fact that the energy used to run this large commuter rail system is generated independent of the grid is energy and efficiency neutral, since the State of New Jersey could have just as easily compelled public utilities to add equal capacity for this necessary service. Additionally, with respect to the reported data, the United States military has committed, for strategic and ecological reasons, to make all domestic military bases energy self-sufficient [46]. Although the adoption of microgrid power consumption by military bases does alleviate the strain on the distribution grid at present, the relief is singular and finite and does not address the future strains which will occur due to increased population densification. In fact, only two reported microgrids in the data set addressed residential users in congested areas. Both are located in Kings County, New York, Brevoort Cogeneration Microgrid, and New York Affordable Housing Microgrid. Both are retrofits, with the structures not optimized to take advantage of the benefits of a microgrid.

It is posited that an integrated, holistic approach to real estate development using multiple technologies in buildings designed to maximize their use is not only socially responsible but also economically viable. Inclusion of the microgrid from the outset would allow buildings within the development to utilize the maximum amount of energy. Therefore, it is proposed that a future development be designed around a grid-connected microgrid capable of island-mode operation as follows:

1. **Main power generation-combined cycle gas and steam plant:** Gas and steam turbines would produce electricity at high efficiency for the development. The waste heat would be used to produce building heat, hot water, and air conditioning.

2. **Flexible sizing of the microgrid:** Depending on local regulations, neighboring entities could also be enlisted into the microgrid. Although not modeled herein, if such entities include vital facilities such as hospitals or fire stations, the microgrid facility may be eligible for non-interruptible status with respect to natural gas supply.

3. **Maximization of renewable energy assets:** Buildings would be designed from the outset to maximize both solar and wind generations, thereby decreasing the carbon footprint of the development overall and the cost of fuel.

4. The development model

A model is presented which incorporates a microgrid utilizing renewable energy assets into a development consisting of three 32-story residential towers and one 57-story commercial tower. The scale of the development is in keeping with current large-scale developments. The model is run in each of the eight cities of interest using local data on environmental conditions, construction costs (exclusive of real estate purchase), and income levels based upon local rents and power rates. Analyses of the benefits of including the microgrid, as opposed to the same scale conventional development, are performed on three levels:

1. **Infrastructure impact:** Electric load is presented with and without the microgrid.

2. **Economic:** Construction cost, anticipated revenue, and time to repay are presented with and without the microgrid.

3. **Ecological:** Greenhouse emissions are compared with and without the microgrid.

4.1 The buildings

Actual hourly electrical usage data and building specifications of one 32-story residential tower and one 57-story commercial tower located in New York City were generously provided by GridMarket LLC, New York City, NY. The dimensions of each tower, given in square feet (sq. ft.) and number of apartments (apts.), are given in **Table 5**. Rentable area for the entire four-tower development is calculated, per industry standard, at 50% of total square footage, with the remaining 50% designated for hallways, stairwell, elevator shafts, and other such types of general-use areas.

	Built	ES	Lot area	Internal area (sq. ft)	Floors	Retail (sq. ft)	Office (sq. ft)	Res. (sq. ft)	Apts.
Res.	1916	94	80,333	938,324	32	26,805	0	879,019	608
Comm.	1972	165	127,966	2,689,635	57	216,912	2,319,046	0	0

Table 5.
Specifications for buildings to be used in the model (provided by GridMarket LLC).

4.2 Power generation

A 2 × 1 (two gas turbines 7.9 MW powering 1 MW steam turbine) was the optimal configuration for the cogeneration plant. This would be supplemented by power provided by 10,500 solar panels and 295 1 kW vertical drum-type wind turbines. The number of solar panels was estimated by covering the entire roof area of the four proposed buildings with standard 77 inch by 39 inch panels, while the number of wind turbines was estimated by placing a turbine every 10 feet around the periphery of each roof. It is recognized that whole-roof coverage with solar panels is impracticable; however, the estimate is valid because some amount of appropriately facing surface area would be available for additional panels. Also, it is assumed for these calculations that the buildings will be boring rectangles. As this is neither likely nor desirable, setbacks will create additional space for more wind turbines. Energy storage devices will be included in the design from the outset in order to balance generated power between times of low load and high load.

4.3 Calculations

4.3.1 Power generation potential

Publically available commercial data was used to estimate all generating capabilities for gas turbines, steam turbines, and wind turbines as follows: two 7.9 kW gas turbines operating at 30.6% efficiency [47] driving a single 750 kW gas turbine [48] raising the total efficiency to 50.2% and 1 kW wind turbines [49]. All power generation was calculated on an hourly basis and balanced with the hourly load as much as possible. Renewable energy sources were given precedence. Annual average daily data for wind speed at 50 meters aboveground [50], sunrise and sunset [51], and average solar irradiance in kWh/m^2 [50] were obtained for each city of interest. As the wind speed and irradiance data were daily averages, they were applied for all 24 hr in each given day. Sunrise and sunset data were used to "turn on" and "turn off" the solar component of the system.

There are multiple methods for determining the efficiency of trigeneration systems [52–55]. For this study, general estimates based upon these methods will be used. The fast-start capability of modern turbines was utilized to estimate cogeneration outputs with one gas turbine operating at all times. If hourly load minus available renewables exceeded the capacity of one gas turbine, the second turbine was started. If the hourly load still exceeded the capacity of both gas turbines, the steam turbine was included. Solar generation was calculated on an hourly basis by multiplying the irradiance by the total panel area (total roof area) at a 15% conversion efficiency and a 75% transmission efficiency. For wind energy, the manufacturer's power generation curve was used [49]. The power curve, with a cut-in at 6 miles per hour of wind speed, was applied to the hourly average wind speed to determine the kW delivered by the posited 295 turbines. Usable waste heat from cogeneration (as well as input fuel needs) was calculated on an hourly basis. Input energy in kW was calculated as hourly output divided by hourly efficiency of the cogeneration set, 30.6% for gas-only generation and 50.2% for combined generation. Gross waste heat was obtained by subtracting this number from generated power (Eq. 1):

$$((\mathrm{kWh/hr})/\mathrm{Efficiency} - (\mathrm{kWh/hr}))/0.0002931\ \mathrm{BTU/kWh} = \mathrm{Gross\ BTU} \quad (1)$$

Usable waste heat was calculated by obtaining the ideal thermodynamic efficiency of the system (Eq. 2) [56] and multiplying this by the results of Eq. 1 (Eq. 3):

$$(T_{high} - T_{low})/T_{high} = 0.61 \quad (2)$$

$$\text{BTU/hour (gross)} \times 0.61 \quad (3)$$

The annual volume of natural gas (NG) required to run the cogeneration system, measured in industry standard cubic feet (ft^3), was calculated by summing the hourly energy input, (kWh/hr) divided by hourly efficiency of cogeneration and converted to cubic feet of gas (Eq. 4):

$$\Sigma(((\text{kWh/hr})/\text{Efficiency}) \times (3412\ \text{Btu/kWh}) \times (1\ \text{ft}^3/103.7\ \text{BTU})) \quad (4)$$

4.3.2 Estimated usage

Building performance in terms of ENERGY STAR rating (EGR) was not modified. Usage was normalized to environmental conditions in each of the given cities as follows. Daily average high and low temperatures for obtained for each city [57]. Hourly temperatures were calculated using a linear regression each day of the year in each city, starting from the daily low for the day at 1:00 a.m. up to the daily high for the day at 12:00 noon and going back to the daily low again at 12:00 midnight. It was assumed that, on any given day, air conditioning (AC) would be required at or above a daily high of 80°F, and heat would be needed at a daily low of 50°F (**Table 6**).

Days	Cairo	Lagos	Shanghai	Mumbai	London	Mexico City	NYC	Sao Paolo
AC	206	365	103	365	0	110	98	245
Heat	63	0	107	0	199	107	178	0
None	96	0	155	0	166	148	89	120

Table 6.
Annual climate control needs for cities of interest.

In order to estimate the base electric usage for the proposed development, the hourly base building usage data provided by GridMarket was increased by 30% for each day for each city that air conditioning was assumed to be needed. (Given that the New York City data is actual usage, days that the data indicated that air conditioning would be needed in both New York and any other given city were not modified.) This provided a reference point for the estimated electrical load for the proposed development being connected to the regional/national power grid. Since inclusion of the microgrid would essentially eliminate electricity usage for air conditioning, daily usage data for the development with an included microgrid was reduced by 30% assuming that air conditioning increases daily load by 30% (**Table 6**).

Hourly heat usage for both hot water and building heat was assumed to remain constant across all cites and climates since the heat capacity of water is constant and the amount of hot water required on a daily basis would be independent of location or climate. Also, as the configuration (and hence the volume) of the buildings was identical in all cities, and the need for heating is temperature dependent, the amount of heat required to provide building heat on an hourly basis would also be constant. Hourly heat requirements for both needs were therefore calculated based upon the New York City data and applied to all cities. The total annual energy use breakdown for New York is available from the United States Energy Information Agency as follows: electricity, 27.2%; heating, 55.8%; and hot water, 17.0% [58].

Total actual electrical usage for the four-tower development was converted to BTUs and divided by 0.272 (27.2%) to give total energy usage (Eq. 5):

$$(1.81 \times 10^8 \text{ kWh Annual Electrical}/.272) \times 3412 \text{ BTU/kWh} = 2.27 \times 10^{12} \text{ BTU/year} \tag{5}$$

This number was then multiplied by 0.17 (17%) to provide annual hot water usage and divided by 8760 hr/year to arrive at an hourly hot water usage of 4.4×10^7 BTU/hr. for hot water (Eq. 6). This was applied to every hour of the year in all cities:

$$(2.27 \times 10^{12} \text{ (BTU/year)} \times 0.17)/8760 \text{ hr./year} = 4.4 \times 10^7 \text{ BTU/hr.hot water} \tag{6}$$

Hourly structural heating power was calculated at 0.558 (55.8%) of total energy divided by actual hours of heat usage in New York. This number of 2.96×10^8 BTU/hour was then applied to each city for each hour (Eqs. 7–8):

$$(2.27 \times 10^{12} \text{ (BTU/year)} \times 0.558) = 1.27 \times 10^{12} \text{ BTU/year building heat} \tag{7}$$

$$1.27 \times 10^{12} \text{ BTU/year}/(178 \text{ days heat/year} \times 24 \text{ hr/day}) = 2.96 \times 10^8 \text{ BTU/hour} \tag{8}$$

Usable hourly waste heat (Eq. 3) was initially applied as needed for climate control. Hourly heat transfer needs were calculated at 80% of available waste heat for building heating, with 80% efficiency being the average efficiency of a standard heat exchanger [59]. Hourly heat transfer needs were calculated at 120% of available waste heat for air-conditioning needs, with 120% being the average efficiency of a two-stage absorption chiller [60]. Remaining heat, on an hourly basis, was then applied to provide hot water, again at 80% efficiency.

5. Results

5.1 Infrastructure impact

Infrastructure impact is defined by the degree that the implementation of an integrated development model would relieve strain on the local power distribution grid. As can be seen in **Table 7**, this is highly correlated to air-conditioning needs as shown in **Table 6**. This is the expected result as heat provided from the trigeneration plant to the absorption chillers replaces electrical load for air conditioning. Mumbai and Lagos, cities which essentially require air conditioning year-round, had the highest reduction in load, while London, which essentially requires no air conditioning, saw no reduction in load.

The microgrid must be grid-connected for both safety and regulatory reasons in an urban environment. To be effective, the system must not add additional load to the grid but must also be balanced in order to protect the local grid infrastructure; it should not push power onto the distribution grid at any point. In case of emergency, such as a blackout condition, the microgrid should also be able to disconnect from the local power grid and provide all needed services in island mode. **Table 7** indicates that the proposed model succeeds in this respect. The incorporated power generating systems produce surplus electricity on an hourly basis between 40% and 60% of the time, depending upon the city ("% hours off-grid"). Excess energy produced in hours of low load can be stored in incorporated batteries to meet demand in hours of high load, producing a system that is completely grid-neutral throughout the year. The annual difference between electricity usage and

City	No microgrid (kWh/year)	Microgrid (kWh/year)	Saved	Microgrid production (kWh/year)	Hours off-grid	Surplus/ deficit (kWh/year)	Surplus/ deficit
Cairo	1.96E+08	1.75E+08	10.6%	1.73E+08	48.0%	(2,274,872)	−1%
Lagos	2.05E+08	1.60E+08	21.8%	1.75E+08	61.8%	14,415,926	9%
Shanghai	1.83E+08	1.80E+08	1.2%	1.76E+08	44.0%	(4,663,087)	−3%
Mumbai	2.16E+08	1.69E+08	21.6%	1.77E+08	57.7%	8,210,049	5%
London	1.81E+08	1.81E+08	0.0%	1.72E+08	41.2%	(9,295,666)	−5%
Mexico City	1.90E+08	1.77E+08	7.0%	1.73E+08	48.3%	(3,766,257)	−2%
NYC	1.81E+08	1.63E+08	10.0%	1.71E+08	55.8%	8,533,701	5%
Sao Paolo	2.12E+08	1.71E+08	19.6%	1.76E+08	51.8%	5,855,500	3%

Table 7.
Load and production comparison (selected cities).

production in each city as shown in **Table 7**, be it positive or negative, is small and can be corrected in the local design phase.

5.2 Socioeconomic impact

The cost of constructing the proposed development buildings was calculated using local average construction cost data and applying it to the total square footage found in **Table 5** (**Table 9**-"Cost-Less Microgrid") [61]. The cost of integrating the microgrid is calculated on average prices in the United States (**Table 8**) [62]. An internally consistent data set containing all cities of interest for this metric was not found. It is assumed that the cost would be fixed since the capital components required are not locally produced. The cost of batteries, absorption chillers, and heat exchangers was not included, as it is assumed that these costs would be balanced by the deletion of HVAC equipment, cooling towers, boilers, and hot water heaters. Cost differential is presented in **Table 9**.

Revenue is calculated on local monthly rental rates per square foot of rental. As previously mentioned, rentable space is calculated at 50% of available floor space. Internally consistent residential rental rates were found [63]. However, internally consistent rates for residential, commercial, and retail were only found for New York City [64, 65]. Therefore, commercial rates for other cities are calculated at the ratio of those rates to residential rates for New York. As the microgrid will also be a revenue source, local electricity rates are included and applied to the revenue for each city [66]. Total annual revenue from rents and electrical for each city, as well as the rates used, is given in **Table 10**. **Table 11** then estimates the gross time to repay the initial investment, with "Cost-Less Microgrid" in **Table 9** divided by the

Cogeneration	**$895**	**Per kW**	**15,660**	**MW**	**$14,015,700**
Solar panels	$2434	Per kW	2100	Watts	$5,111,400
Wind turbines	$1630	Per kW	292	Turbines	$475,960
				Total	**$19,603,060**

Table 8.
Microgrid cost.

City	Res. twrs	Com. tws	Total sq. ft.	Per sq. ft. res.	Per sq. ft. com.	Cost-less microgrid	Cost-with microgrid
Cairo	3	1	5,504,607	$31.13	$35.94	$184,295,560	$203,898,620
Lagos	3	1	5,504,607	$30.00	$32.00	$170,517,480	$190,120,540
Shanghai	3	1	5,504,607	$21.23	$33.82	$150,725,311	$170,328,371
Mumbai	3	1	5,504,607	$18.96	$20.09	$107,406,636	$127,009,696
London	3	1	5,504,607	$112.63	$120.56	$641,312,692	$660,915,752
Mexico City	3	1	5,504,607	$52.41	$22.36	$207,672,921	$227,275,981
NYC	3	1	5,504,607	$285.32	$534.00	$2,239,432,901	$2,259,035,961
Sao Paolo	3	1	5,504,607	$18.11	$42.17	$164,401,051	$184,004,111

Table 9.
Construction cost comparison of proposed development: incorporating vs. not incorporating a microgrid.

sums of all rentals in **Table 10** to determine the number of years to repay the development if built conventionally and the "Cost-With Microgrid" divided by the sum of all rentals plus electricity revenue in **Table 10** used. Operating and real estate costs were not considered in the gross time to repay, but it can be assumed that these costs will be identical in both scenarios in any given city.

From **Table 11**, it can be seen that the gross time to repay initial investment is lower when the microgrid is present. This has positive sociological implications. Since repayment time is shorter, long-term revenue will be higher, making the proposed development model economically profitable and therefore feasible. This has an additional advantage; the charging of premium rents is not economically required due to the lower repayment time of a microgrid inclusive development. The enhanced revenue stream and lowered operating costs associated with building a development around this model would also make affordable housing economically viable, serving to include those who are often left behind and displaced when a neighborhood is redeveloped.

5.3 Environmental impact

Buildings generate greenhouse gases indirectly by consuming electricity produced from various fuels and directly generate such gases through the production of heat and hot water. **Table 12** presents the breakdown of fuels used to generate electricity for the local power grid in each of the cities of interest [67]. **Table 13** shows the greenhouse gas emissions for each of those sources per kWh [68]. **Table 14** presents the percentage of energy derived from renewable sources incorporated into the microgrid in each city of interest. **Table 15** contains data on natural gas usage for the development both with and without inclusion of the microgrid. In both cases, annual hot water needs are calculated according to Eq. 7 multiplied by 8760 hr/year, and heating needs are calculated by Eq. 8 multiplied by the number of heating hours estimated in each city from **Table 6**. For the traditional version of the development, it is assumed that these needs will be supplied by burning natural gas, although less environmentally friendly fuel oil could also be used. When the microgrid is present, heat and hot water needs are met first by trigeneration waste heat, and any unmet needs are met by the same natural gas feed that would fuel the gas turbines. Finally, **Table 16** compares the calculated greenhouse gas emissions between the two scenarios with data from **Tables 8, 13, 14** and **15**.

City	Res./sq. ft./month	Office/sq. ft./month	Retail/sq. ft./month	Electricity/kWh	Res. ent/year	Office rent/year	Retail rent/year	Electricity/year
Cairo	$1.11	$2.35	$17.51	$0.02	$17,562,800	$32,725,167	$31,235,468	$3,501,974
Lagos	$1.12	$2.37	$17.67	$0.08	$17,721,023	$33,019,989	$31,516,869	$3,207,401
Shanghai	$1.07	$2.27	$16.88	$0.09	$16,929,906	$31,545,882	$30,109,866	$3,607,725
Mumbai	$0.60	$1.27	$9.46	$0.07	$9,493,405	$17,689,280	$16,884,037	$3,382,754
London	$2.40	$5.09	$37.86	$0.22	$37,973,621	$70,757,118	$67,536,148	$3,617,546
Mexico City	$0.49	$1.04	$7.73	$0.08	$7,752,948	$14,446,245	$13,788,630	$3,536,433
NYC	$3.45	$7.31	$54.42	$0.18	$54,587,080	$101,713,358	$97,083,212	$3,425,333
Sao Paolo	$0.53	$1.12	$8.36	$0.19	$8,385,841	$15,625,530	$14,914,233	$3,412,530

Table 10.
Estimated rental and electrical rates and annual revenues per source.

City	Oil	NG	Coal	Nuclear	Hydroelectric	Non-hydroelectric renewables
Cairo	44.67%	50.72%	0.47%	0.00%	3.49%	0.65%
Lagos	42.12%	28.25%	21.80%	0.82%	5.87%	1.14%
Shanghai	18.95%	6.20%	61.83%	1.58%	8.62%	2.82%
Mumbai	29.38%	6.23%	56.91%	1.18%	4.03%	2.27%
London	38.89%	36.70%	5.83%	8.63%	0.65%	9.31%
Mexico City	44.41%	43.20%	5.26%	1.28%	3.63%	2.21%
NYC	0.00%	44.00%	1.00%	31.00%	19.00%	5.00%
Sao Paolo	46.61%	11.06%	5.55%	1.21%	29.19%	6.38%

Table 11.
Electric power generation source fuels.

City	Oil	NG	Coal	Nuclear	Hydroelectric	Nonhydroelectric renewables
Cairo	44.67%	50.72%	0.47%	0.00%	3.49%	0.65%
Lagos	42.12%	28.25%	21.80%	0.82%	5.87%	1.14%
Shanghai	18.95%	6.20%	61.83%	1.58%	8.62%	2.82%
Mumbai	29.38%	6.23%	56.91%	1.18%	4.03%	2.27%
London	38.89%	36.70%	5.83%	8.63%	0.65%	9.31%
Mexico City	44.41%	43.20%	5.26%	1.28%	3.63%	2.21%
NYC	0.00%	44.00%	1.00%	31.00%	19.00%	5.00%
Sao Paolo	46.61%	11.06%	5.55%	1.21%	29.19%	6.38%

Table 12.
Electric power generation source fuels.

Lbs/BTU	Coal	Oil	NG	Solar	Hydroelectric	Nuclear	Wind
CO_2	2.15E-04	1.61E-04	1.17E-04	2.89E-05	1.54E-05	7.70E-06	7.10E-06
SO_2	2.59E-06	1.12E-06	7.00E-09	*Negligible*	*Negligible*	*Negligible*	*Negligible*

Table 13.
Greenhouse gas emissions per source fuel.

KwH/year	Cairo	Lagos	Shanghai	Mumbai	London	Mexico City	NYC	Sao Paolo
Total	1.73E+08	1.75E+08	1.76E+08	1.77E+08	1.72E+08	1.73E+08	1.71E+08	1.76E+08
Wind	2.47E+06	7.62E+05	2.17E+06	1.74E+06	2.06E+06	1.70E+06	2.26E+06	2.01E+06
Solar	2.89E+07	3.35E+07	3.18E+07	3.48E +07	2.76E+07	2.99E+07	2.81E+07	3.32E+07
Renewable	**18.1%**	**19.6%**	**19.3%**	**20.6%**	**17.3%**	**18.3%**	**17.7%**	**20.0%**

Table 14.
Percent renewable power generation on microgrid.

	On national power grid				Trigenerating microgrid					
	Cubic feet NG annually				Cubic feet NG annually					Difference
City	Hot water	Heat	AC	SUM	Hot water	Heat	AC	Turbine	SUM	
Cairo	1.0E+07	4.3E+09	0	4.4E+09	0.0E+00	2.7E+09	0.0	1.5E+10	1.8E+10	**1.4E+10**
Lagos	1.0E+07	0.0E+00	0	1.0E+07	0.0E+00	0.0E+00	0.0	1.5E+10	1.5E+10	**1.5E+10**
Shanghai	1.0E+07	7.4E+09	0	7.4E+09	0.0E+00	4.5E+09	0.0	1.5E+10	2.0E+10	**1.2E+10**
Mumbai	1.0E+07	0.0E+00	0	1.0E+07	0.0E+00	0.0E+00	0.0	1.5E+10	1.5E+10	**1.5E+10**
London	1.0E+07	1.4E+10	0	1.4E+10	1.0E+07	9.5E+09	0.0	1.3E+10	2.2E+10	**8.8E+09**
Mexico City	1.0E+07	7.3E+09	0	7.3E+09	0.0E+00	4.6E+09	0.0	1.5E+10	2.0E+10	**1.2E+10**
NYC	1.0E+07	1.2E+10	0	1.2E+10	1.0E+07	7.8E+09	0.0	1.5E+10	2.3E+10	**1.0E+10**
Sao Paolo	1.0E+07	0.0E+00	0	1.0E+07	0.0E+00	0.0E+00	0.0	1.5E+10	1.5E+10	**1.5E+10**

Table 15.
NG consumption: with microgrid compared to without microgrid.

These results are significant. As seen in **Table 15**, a development incorporating a microgrid uses over 10 times the natural gas in all cases than the identical development drawing power from the local distribution grid. However, **Table 16** definitively shows that the use of a microgrid would greatly reduce the greenhouse gas emissions from the development, with approximately half of the CO_2 and virtually all SO_2 emissions eliminated. By incorporating trigeneration from the outset, all upstream emissions from electricity generation are eliminated. Additionally, the use of waste heat in the building systems eliminates emissions from the production of hot water, halves the emissions from building heat, and also eliminates any emissions from air conditioning (bearing in mind that, in a conventional arrangement, air-conditioning emissions would be included in electricity generation emissions). Finally, **Table 14** shows that incorporating maximal renewable assets by design accounts for roughly 20% of the electricity production which, at 50% generator efficiency,

Tons/year	On power grid		Microgrid		Reduction (%)	
City	CO_2	SO_2	CO_2	SO_2	CO_2	SO_2
Cairo	9.84E+04	4.80E+02	4.46E+04	2.67E+00	54.64%	99.44%
Lagos	7.54E+04	5.25E+02	2.81E+04	1.68E+00	62.80%	99.68%
Shanghai	1.12E+05	6.56E+02	5.58E+04	3.34E+00	50.08%	99.49%
Mumbai	7.93E+04	6.73E+02	2.81E+04	1.68E+00	64.55%	99.75%
London	1.49E+05	4.93E+02	8.58E+04	5.14E+00	42.53%	98.96%
Mexico City	1.14E+05	4.70E+02	5.62E+04	3.36E+00	50.88%	99.28%
NYC	1.41E+05	8.04E+02	7.56E+04	4.52E+00	46.21%	99.44%
Sao Paolo	7.81E+04	5.01E+02	2.82E+04	1.69E+00	63.87%	99.66%

Table 16.
Comparison of greenhouse emissions for development: without vs. with microgrid.

amounts to a 40% drop in potential greenhouse gas production through electricity generation.

6. Conclusions

Urbanization of populations is occurring at an accelerating pace worldwide, and, in all countries, the increasing densification of population is putting a strain on the pre-existing infrastructure. Depending on the state of national economic development, that infrastructure could be robust, aging, or nonexistent, but was not designed to support the increasing strain. Additionally, this seismic population shift requires housing and employment opportunities in relatively small geographic areas. While growth has always bought opportunity, that opportunity was never immediate or evenly distributed. Hence, slum populations are increasing, and both housing and economic opportunity are increasingly scarce.

History has shown that economies cannot be managed, but it is the job of the government to "promote the general welfare" [69]. At present, various local, national, and international entities are promoting the general welfare through establishing programs to create sociological and environmentally sustainable opportunity. These incentives recognize the existence of a need which can be addressed by a new model of urban growth, one that is economically advantageous and sociologically and environmentally sound, such as the design model proposed herein. The work presented develops a new model for urban development, a model which incorporates a myriad of mature technologies into a real estate development at the design stage. The model bases the development around a self-contained microgrid using trigeneration of power where, at the first stage, fossil fuel-powered turbines produce electricity and heat which, at the second stage, powers a steam turbine to produce more electricity. The third stage of trigeneration is to use the remaining exhaust heat to provide building heat, hot water, and air conditioning. The system is supplemented by renewable solar and wind power, with the buildings designed from the outset to maximize such assets. Modern power storage assets are included in the design to balance the load between times of high usage and low usage.

The study demonstrates that the proposed model succeeds in meeting all sustainability requirements. It is more profitable than constructing the same development on the national power grid. This is vital since economic sustainability is a sine qua non for any urban development. It balances load and generation capability on a large scale, allowing the construction of large numbers of buildings to accommodate increasing populations with essentially no impact on the existing power distribution infrastructure. It is also environmentally sustainable, producing fewer emissions than traditional developments on the same scale.

These conclusions point to the viability and the economic and environmental desirability of proceeding with urban development under the model herein presented and also lead to a sociological conclusion. Cities are historically built by the poor striving to make a better life for themselves and their families. In developed countries, the consequence of real estate development is too often to push such people out of their homes and further to the fringes. In developing countries, such people are often not even considered, relegated to living in shanty towns. The economic and environmental advantages of this development model present an opportunity to promote the general welfare of all. Environmental financial incentives, coupled with increased profitability, will allow for the maintenance of exceptional living conditions at comparatively low rents. Since renewability and regeneration are incorporated into the building design, heat and hot water, so necessary for everyday life, will be readily available. Lower rents are economically possible since all these usual living expenses are being provided for by the same source. Most importantly, the model is scalable and variable and power is fungible. A commercial tower was included in this study to both provide an economic focus point to start the development and to provide jobs to the people living in the residential towers because real-world data was made available. The model could be applied to any combination of residential, commercial, retail, or industrial spaces providing centers for human advancement, with both jobs and housing provided at an economically and environmentally favorable rate.

Acknowledgements

The author would like to acknowledge the assistance of the following people:

- Mr. Richard Denis—*Siemens Energy, Siemens AG*—for providing details on numerous power generation combinations
- Mr. Oisín O'Brien—*GridMarket LLC*—for providing real-world usage data
- Mr. William Kenworthey—*Hellmuth, Obata and Kassabaum (HOK), New York, New York*—for the advice on practicality and design
- Mr. Anthony Alduino—*Thornton Tomasetti LLC*—for providing working information on integrating microgrids into urban environments
- Mr. Bradford Sussman-Gonzalez—*Pitta Bishop & Del Giorno LLC, New York, New York*—for providing practical insight on regulatory issues

Dedication

This work is dedicated to the memory of Mr. Peter Tymus, formerly of Turner Constructions, Inc. and Long Island University, Brooklyn, NY. Mr. Tymus was a gentleman in the truest sense of the word, and his vision and sharing of knowledge were the inspiration for this work.

DOI: http://dx.doi.org/10.5772/intechopen.83560

Author details

Edward J. Donahue
Paramount Technical Consulting, Brooklyn, NY, USA

*Address all correspondence to: ejdonahue1965@yahoo.com

References

[1] International Bank for Reconstruction and Development—Commission on Growth and Development, Urbanization and Growth, Spence M, Annez PC, Buckley RM, editors. Washington, DC: The World Bank; 209, pp. 1-45

[2] Davis K. The Origin and Growth of Urbanization in the World. American Journal of Sociology. 1955;**60**(5): 429-437

[3] Bloom D, Canning GFD. Urbanization and the Wealth of Nations. Science. 2008;**319**(5864): 772-775

[4] Narayan L. Urbanization and Development. International Journal of Research. 2014;**1**(8):901-908

[5] Ha Minh Nguyen LDN. The relationship between urbanization and economic growth: An empirical study on ASEAN countries. International Journal of Social Economics. 2018;**45**(2): 316-339

[6] Zhao Y, Wang S. The relationship between urbanization, economic growth and energy consumption in China: An econometric perspective analysis. Sustainability. 2015;**7**(5):5609-5627

[7] Szreter S. Industrialization and health. British Medical Bulletin. 2004; **69**(1):75-86

[8] World Bank Group. Data Bank: World Bank Development Indicators [Internet]. Available from: http://databank.worldbank.org/data/reports.aspx?source=2&series=SP.URB.TOTL# [Accessed: 27-10-2018]

[9] World Bank Group. Data Bank: World Bank Development Indicators [Internet]. Available from: http://api.worldbank.org/v2/en/indicator/SP.POP.TOTL?downloadformat=excel [Accessed: 27-10-2018]

[10] Fielding JE. Public health in the twentieth century: Advances and challenges. Annual Review of Public Health. 1999;**20**(1999):xxii-xxxx

[11] Ivan Torok GM. Urbanization and economic growth: The arguments and evidence for Africa and Asia. Environment and Urbanization. 2013; **25**(2):465-482

[12] The International Bank for Reconstruction and Development. The State of Electricity Access. Washington, DC: The World Bank; 2017

[13] International Energy Agency. Sustainability Goal 7: Ensure Access to Affordable, Reliable, Sustainable and Modern Energy for All. Office of Economic Cooperation and Development [Internet]. 1 March 2018. Available: https://www.iea.org/sdg/electricity/ [Accessed: 3-11-2018]

[14] USDOE. Smart Grid R&D Program. DOE Microgrid Workshop Report. August 2012. United States Department of Energy Office of Electricity and Energy Reliability. 2012

[15] United Nations, Department of Economic and Social Affairs, Population Division. The World's Cities 2016. New York: United Nations; 2016

[16] Smith D. World City Populations: 1950–2036. citygeographics.org [Internet]. 2018. Available from: http://luminocity3d.org/WorldCity/#6/49.767/26.642 [Accessed: 3-10-2018]

[17] United Nations Department of Economic and Social Affairs Population

Division. World Urbanization Prospects. United Nations Department of Economic and Social Affairs [Internet]. 2018. Available from: https://population.un.org/wup/Download/Files/WUP2018-F12-Cities_Over_300K.xls [Accessed: 4-11-2018]

[18] United Nations Department of Economic and Social Affairs: Population Division. 2017 Revision of World Population Prospects. United Nations [Internet]. 2018. Available from: https://population.un.org/wpp/ [Accessed: 10-11-2018]

[19] Canty and Associates LLC. Weatherbase. CantyMedia [Internet]. Available from: https://www.weatherbase.com/ [Accessed: 10-11-2018]

[20] Arnfield J. Encyclopedia Britannica. Encyclopedia Britannica, Inc. [Internet]. 25 October 2015. Available from: https://www.britannica.com/science/Koppen-climate-classification [Accessed: 11-11-2018]

[21] United Nations Development Programme. Human Development Reports: The Human Development Index (HDI). New York: United Nations; 2018

[22] United Nations Statistics Division-Department of Economic and Social Affairs. Millenium Development Goal Indicators. United Nations [Internet]. 2018. Available from: http://mdgs.un.org/unsd/mdg/Data.aspx [Accessed: 17-11-2018]

[23] United Nations Human Settlement Program. UN_HABITAT: State of the World's Cities 2006/7. United Nations [Internet]. 2008. Available from: http://mirror.unhabitat.org/documents/media_centre/sowcr2006/SOWCR%205.pdf [Accessed: 11-11-2017]

[24] The Borgen Project. Poverty in Cairo. The Borgen Project [Internet]. 2018. Available from: https://borgenproject.org/poverty-cairo/ [Accessed: 3-11-2018]

[25] Word Population Review. World Population Review-Lagos 2018. World Population Review [Internet]. 2018. Available from: http://worldpopulationreview.com/world-cities/lagos-population/ [Accessed: 3-11-2018]

[26] World Population Review. Mumbai Population 2018. World Population Review [Internet]. 2018. Available from: http://worldpopulationreview.com/world-cities/mumbai-population/ [Accessed: 3-11-2018]

[27] The Greater London Authority. London Data Store-Poverty in London 2015/16. The Greater London Authority [Internet]. 2018. Available from: https://data.london.gov.uk/apps_and_analysis/poverty-in-london-201516-2/ [Accessed: 3-11-2017]

[28] The Borgen Project. Poverty in Mexico City Exists. The Borgen Project [Internet]. 2018. Available from: https://borgenproject.org/poverty-in-mexico-city/ [Accessed: 3-11-2018]

[29] The United States Census Bureau. Quick Facts: New York City, New York. United States Department of Commerce [Internet]. 2018. Available from: https://www.census.gov/quickfacts/fact/table/newyorkcitynewyork/PST045217 [Accessed: 3-11-2018]

[30] The Borgen Project. Revamping Favelas: Top 10 Facts About Poverty in Sao Paulo. The Borgen Project [Internet]. 2018. Available from: https://borgenproject.org/tag/slums-in-sao-paulo/ [Accessed: 3-11-2018]

[31] The World Bank. Access to Electricity (% of Population). The World Bank Group [Internet]. 2018. Available from: https://data.worldbank.

org/indicator/EG.ELC.ACCS.ZS [Accessed: 27-11-2018]

[32] The World Bank Group. Enterprise Surveys: What Businesses Experience-Country Profiles. The World Bank Group [Internet]. 2014. Available from: http://www.enterprisesurveys.org/reports [Accessed: 27-10-2018]

[33] The World Bank Group. Data Tables: Section II-Infrastructure, 2.05 Quality of Electrical Supply. The World Bank Group[Internet]. 2018. Available from: http://siteresources.worldbank.org/INTEXPCOMNET/Resources/2.05_Quality_of_Electricity_Supply.pdf [Accessed: 27-10-2018]

[34] The World Bank. TCdata360-Getting Electricity: Reliability of Supply and Transparency of Tariffs. The World Bank [Internet]. 2018. Available from: https://tcdata360.worldbank.org/indicators/hc6a71595?country=BRA&indicator=422&viz=line_chart&years=2014,2017 [Accessed: 27-10-2018]

[35] Arlet J, Hrytskevick V, Mortada H, Parvanyan T, Srinlvasan J, Tjong E. Doing Business 2016. Washington, DC: The World Bank; 2016

[36] The World Bank Group. Electric Power Transmission and Distribution Losses (% of Output). The World Bank Group [Internet]. 2018. Available from: https://data.worldbank.org/indicator/EG.ELC.LOSS.ZS [Accessed: 27-10-2018]

[37] Energypedia. Egypt Energy Situation. Energypedia [Internet]. 2018. Available from: https://energypedia.info/wiki/Egypt_Energy_Situation#cite_ref-Atlam.2C_B._and_Rapiea.2C_A._.282016.29._Assessing_the_Futur [Accessed: 17-11-2018]

[38] Ayokunle OS. The Erratic Electricity Power Supply in Nigeria: Causes and Remedy. Akure: Federal University of Technology; 2015

[39] Science and Technology Select Committee. The Resilience of the Electricity System: HL Paper 121. London: The Stationary Office Limited; 2015

[40] International Energy Agency. Energy Policies Beyond IEA Countries: Mexico 2017. Paris: Interantional Energy Agency; 2017

[41] United States Department of Energy. Grid 2030–A National Vision for Electricity's Second 100 Years. July 2003

[42] Coelho ST, Goldemberg J. Energy access: Lessons learned in Brazil and perspectives for replication in other developing countries. Energy Policy. 2013;**61**:1088-1096

[43] Jai S. Why Mumbai Gets 24×7 Power Supply and Other Cities Don't. Business Standard Private Ltd. [Internet]. 8 September 2015. Available from: https://www.business-standard.com/article/opinion/why-mumbai-gets-24x7-power-supply-and-other-cities-don-t-115090400308_1.html [Accessed: 17-11-2018]

[44] Yang N, Yuan Q, Li L, Ma H, Lin L, Zhu S. Comparison and analysis of metropolitan power grid partition in China and abroad. In: 3rd International Conference on Electric and Electronics (EEIC 2013). Hong Kong; 2013

[45] Microgrid Projects. Microgrid Projects[Internet]. 2017. Available from: http://www.microgridprojects.com [Accessed: 30-06-2017]

[46] Andrews A. Department of Defense Facilities Energy Conservation Policies and Spending. Congressional Research Service. 2009

[47] Siemens AG. SGT-300 Industrial Gas Turbine. Siemens AG [Internet]. 2018. Available from: https://www.siemens.com/global/en/home/products/

energy/power-generation/gas-turbines/sgt-300.html [Accessed 23-09-2018]

[48] Siemens AG. Dresser Rand Steam Turbines. Siemens AG [Internet]. 2018. Available from: https://www.siemens.com/global/en/home/products/energy/power-generation/steam-turbines/d-r-steam-turbines.html [Accessed: 23-09-2018]

[49] Mariah Energy. Windspire Wind Turbines. Mariah Energy [Internet]. 2018. Available from: https://www.windspireenergy.com [Accessed: 15-09-2018]

[50] United States National Aeronautic and Space Agency. POWER Project Data Set. NASA Prediction of World Energy Resources [Internet]. 2018. Available from: https://power.larc.nasa.gov/data-access-viewer/ [Accessed: 22-10-2018]

[51] Edwards Apps, Inc. Sunrise Sunset Cal endars. Edwards Apps, Inc. [Internet]. 2018. Available from: https://www.sunrisesunset.com/ [Accessed: 10-11-2018]

[52] Bilge D, Temir G. Thermoeconomic analysis of a trigeneration system. Applied Thermal Engineering. 2004;**24**: 2689-2699

[53] Hernandez-Santoyo J, Sanchez-Cifuentes A. Trigeneration: An alternative for energy savings. Applied Energy. 2003;**76**:219-227

[54] Kavvadia K, Maroulis Z. Multi-objective optimization of a trigeneration plan. Energy Policy. 2010;**28**:945-954

[55] Khaliq A. Exergy analysis of gas turbine trigeneration system for combined production of power, heat and refrigeration. International Journal of Refrigeration. 2003;**32**:534-545

[56] Levine I. Physical Chemistry, 6th Edition. New York: McGraw-Hill; 2015

[57] The Weather Company. Intellicast Local Interactive. TWC Product and Technology LLC [Internet]. 2018. Available from: http://www.intellicast.com [Accessed: 6-11-2018]

[58] United States Energy Information Agency. Residential Energy Consumption Survey [Internet]. 2009. Available from: https://www.eia.gov/consumption/residential/data/2009/index.php?view=consumption [Accessed: 7-07-2017]

[59] Fakheri A. Heat exchanger efficiency. Journal of Heat Transfer. 2007;**129**(9):1268-1276

[60] United States Department of Energy. Combined Heat and Power Technology Fact Sheet Series: Absorption Chillers for CHP Systems. Washington DC: United States Department of Energy; 2017

[61] Turner and Townsend LLC. International Construction Market Survey 2016. Chicago, Illinois, USA: Turner and Townsend LLC; 2016

[62] United States Energy Information Administration. Construction Cost Data for Electric Generators Installed in 2016. United States Energy Information Administration [Internet]. 2016. Available from: https://www.eia.gov/electricity/generatorcosts/ [Accessed: 20-10-2018]

[63] Numbeo. Cost of Living. Numbeo [Inetrnet]. 2018. Available from: https://www.numbeo.com/cost-of-living [Accessed: 8-10-2018]

[64] La Guerre L. Manhattan Asking Retail Rents Plummet 20 Percent: Report. The Commercial Observer [Internet]. 26 April 2018. Available from: https://commercialobserver.com/2018/04/manhattan-asking-retail-rents-plummet-20-percent-report/ [Accessed: 25-10-2018]

[65] Mashayekhi R. Downtown Office Asking Rents Reach All-Time High: CBRE. The Commercial Observer [Online]. 29 June 2017. Available from: https://commercialobserver.com/2017/06/downtown-office-asking-rents-reach-all-time-high-cbre/ [Accessed: 25-10-2018]

[66] Global Energy Prices.Com. Electricity Prices. Global Energy Prices. Com [Internet]. 2018. Available from: https://www.globalenergyprices.com/en/electricity-prices/ [Accessed: 11-10-2018]

[67] Ritchie H, Roser M. Energy Production and Changing Energy Sources. Ourworldindata.org [Internet]. 2018. Available from: https://ourworldindata.org/energy-production-and-changing-energy-sources [Accessed: 4-10-2018]

[68] United States Environmental Protection Agency. United States Environmental Protection Agency [Internet]. 2016. Available from: https://www.epa.gov/energy/emissions-generation-resource-integrated-database-egrid [Accessed: 27-06-2017]

[69] Constitutional Convention of the United States. The Constitution of the United States of America. Philadelphia, Pensylvannia: United States of America; 1789

Chapter 2

Microgrid Application in Algeria Saharian Remote Areas

Mounir Khiat, Sid Ahmed Khiat, Mohamed Mankour and Leila Ghomri

Abstract

This paper presents a model and simulation for the development of microgrids in remote areas of the Algerian Sahara, including micro power plants, photovoltaic panels, wind farms, diesel energy and storage facilities. The climate of the Algerian Sahara, located on both sides of a tropical region, is hot, sunny and arid. Daytime temperatures are very high and can exceed 50°C, while the thermal amplitude between day and night is often above 350 or 400°C. In addition, there are many microclimates that are characterised by very high wind speeds. This means that wind energy and photovoltaic energy are both widely appropriate in this field, especially if we assume that the distribution of the population is very dispersed. The creation of microgrids for consumption will be an interesting solution to provide energy to the local population. The microgrid is part of the electrical system and is very dynamic. Production and supply forecasts will lead to reshipment, demand and price effects on regional markets. These feedback effects must be modelled and understood to achieve a stable energy system based on renewable energy.

Keywords: remote areas, design microgrid, modelling, real-time simulation

1. Introduction

In the south of Algeria, temperature passes through two extreme values, i.e., from −10 to 34°C (14–93°F), and in some years can reach 49°C (120°F). We see that temperature variation in a day can be more than 44°C (80°F).

On the other hand, winds in the same area are frequent and strong. Rainfall is distributed in irregular manner along the year [1, 2].

The Sahara region is divided into several areas, such as, Bas-Sahara [1], which represents more than half of the population; the western and southern extremities, which are still sparsely populated (less than 0.2 h/km^2), as shown in **Table 1**; followed by Central Western Sahara, where four regions have just over 500,000 inhabitants; and, the northeast Sahara, which represents the most populated region of the Sahara: 1,900,000 inhabitants, or more than two thirds of the Saharan population, live in an area of less than 10% of the desert, with an average density of more than 10 inhabitants/km^2 [3].

These data show that it is impossible to supply energy from conventional power plants. This is why we propose a microgrid model that will be autonomous and self-controlled to ensure continuity of service.

Microgrid power supply networks are emerging to produce, distribute and regulate the flow of electricity at the local level. Microgrids are ideal for university

Power	Realisation			Evolution rate (%)
Maximal power needed from interconnected network	2014	2015	2016	
	10927	12380	28390	3.7
Maximal power from Adrar pole (kw)	261	279	288	8.2
Maximal power from isolated south network (kw)	214	227	242	25.5

Table 1.
Evolution of forecasting power demand.

campuses, military bases, remote sites, office buildings and industrial sites. Improving power quality and reducing transmission losses, robustness and resilience are the main characteristics of microgrid systems [4].

The electricity market in Algeria is very important. Sonelgaz (National Society for Electricity and Gas) is the main supplier of electricity and gas. The country's power plants are open-cycle gas turbines, combined-cycle gas turbines, conventional steam turbines and, more recently, renewable energy sources. Recently, the coverage capacity of the electrical installation's network has reached 98%, with more than 80% in the north [5].

The objective of the real-time simulation is to test the various electrical equipments, under the most natural conditions possible: as if they were connected to the real physical systems associated with them.

The results obtained by this tool will allow us to have a very precise vision of the functioning of microgrids, in terms of power flow or default responses.

The remaining part of the paper is organised as follows: Section 2 describes the evolution of renewable energies in Algeria. In Section 3, we discussed the modelling of a microgrid in Adrar area, located southwest of Algiers, and Section 4 presents a real-time simulation. Finally, Section 5 summarises the results of this work and perspectives of future works.

2. Renewable energy evolution in Algeria

Algeria's national priority is to achieve the multiplication of energy produced by renewable sources. Algeria has great potential in solar energy and may become a leader in the MENA region, and an interesting partner in the world.

Solar and wind energies tend to have become a more serious and efficient solution to enhance supply security and reliability of the whole power system. It contributes too to increasing energy conversion efficiency, transmission and distribution [6].

Since the 1970s, the national programme has strongly supported the use of solar and wind energies; solar is positioned in the first position, but now wind power production is also being strengthened, as shown in **Figure 1**.

Economically, conventional power supply by extending the networks is not suitable for remote centres. This is true for the Sahara regions with its surface area of about 2 million km^2, and only an autonomous mean of supply is to be considered. For this reason, microgrid is the appropriate solution.

Table 1, published by the Algerian National Society for Electricity and Gas (SONELGAZ), shows the evolution of the power forecasts in southern Algeria and in particular in the Adrar area studied.

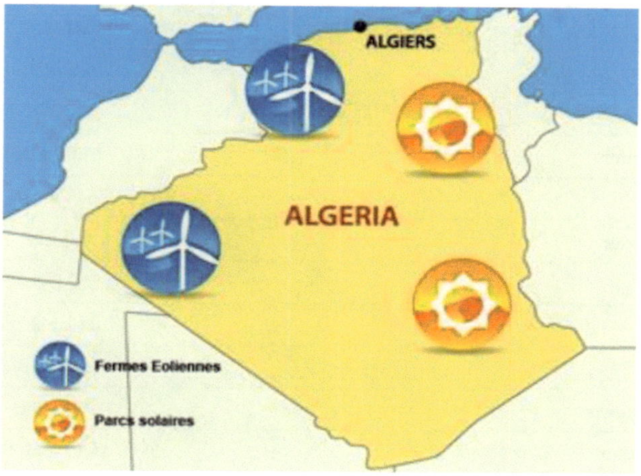

Figure 1.
Algerian renewable energy distribution.

3. Modelling a microgrid in the Adrar study area

Microgrids must adapt existing frameworks. For this reason, the project must be carefully coordinated between the proponent, the utility and the proponent's energy advisor to ensure the best possible result.

For electricity interconnection with the grid and to optimise reliability, resilience and efficiency of the grid, utility tariffs and franchise fees must be adapted.

When designing the microgrid stage, the interconnection specifications of local utilities must be taken into account.

3.1 General architecture

In the first step to designing our microgrid, we will begin by the wind farm of Kabertene (70 km north of Adrar), which produce about 10 megawatts (MW).

This parc is composed of 12 wind turbines, each one have a unit capacity of 0.85 MW. It is interconnected with the 220/30 KV substation located in the locality itself [10].

A simulation model is presented in **Figure 2** (**Table 2**).

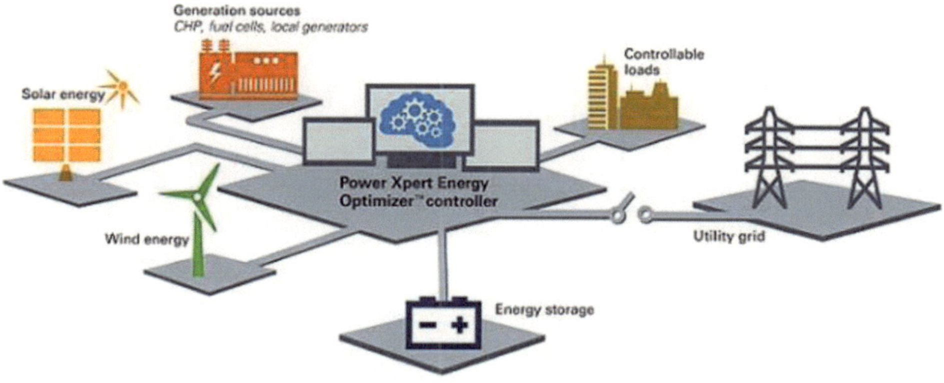

Figure 2.
Microgrids architecture.

Society	GAMESA
Model	G52/850
Wind turbine number	12
Nominal power/unit	850.0 kW
Energy per year	3.42 GWh (Vmoy: 8.5 m/s)
Secondary generator	0.0 kW
Rotor diameter	52.0 m
Column	Tubular shape
Generator type	Variable
Rms in nominal power	26.2 t/mn
Height of the column	55.0; 44.0; 49.0; 65.0 m

Table 2.
Characteristics of wind farm.

The Adrar region seems to be the best location for the installation of wind farms. As shown in **Figure 3**, the value of wind speed is proportional to wind energy production [10].

Several studies have assumed that the average wind speed is about 14 m/s, which is an interesting value [1].

In optimal conditions, power produced by wind turbine is about 10 MW as we see in **Figure 4** below.

The second component of our microgrid is a power plant with eight gas turbines located at "Zaouiet Kounta" (80 km from Adrar). This unity produces 148 MW and allows the Wilaya (department) to ensure its self-sufficiency over 10 years and contribute to supply the city with two power lines, one in the south towards In Salah (Tamanrasset) and the other in the north towards Timimoun (Adrar).

The third energy source is the Kabertène solar power plant which can reach (3 MW) in optimal operation conditions.

The power produced by this photovoltaic source is represented in **Figure 5**.

In comparison with wind energy which is quite stable, solar energy can be used from 7 am to 19 pm (**Figure 6**).

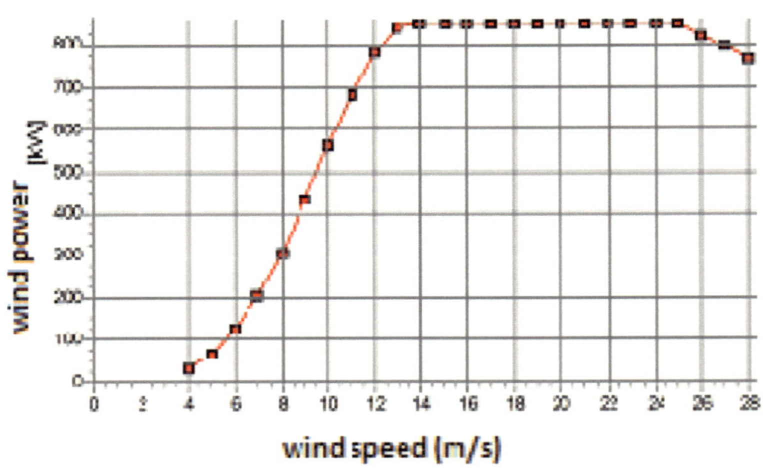

Figure 3.
Wind speed curve.

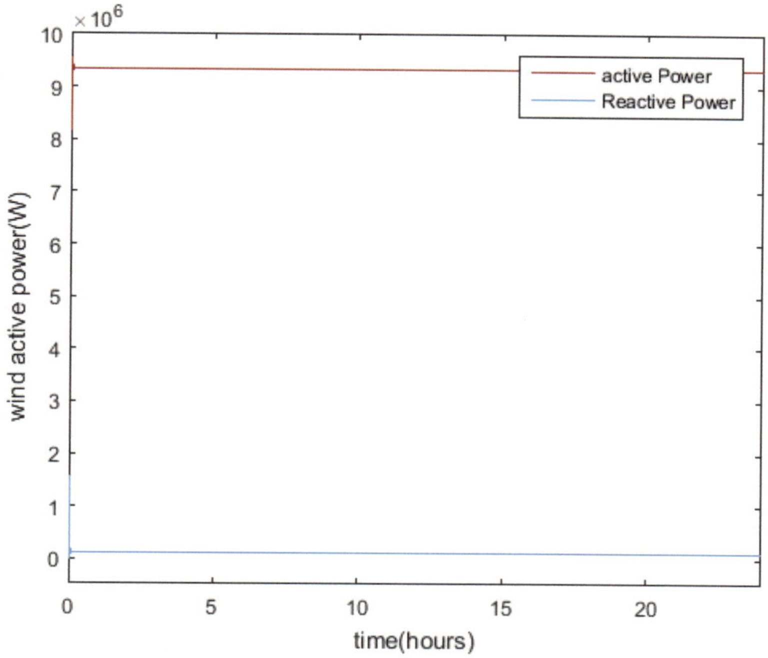

Figure 4.
Wind power curve.

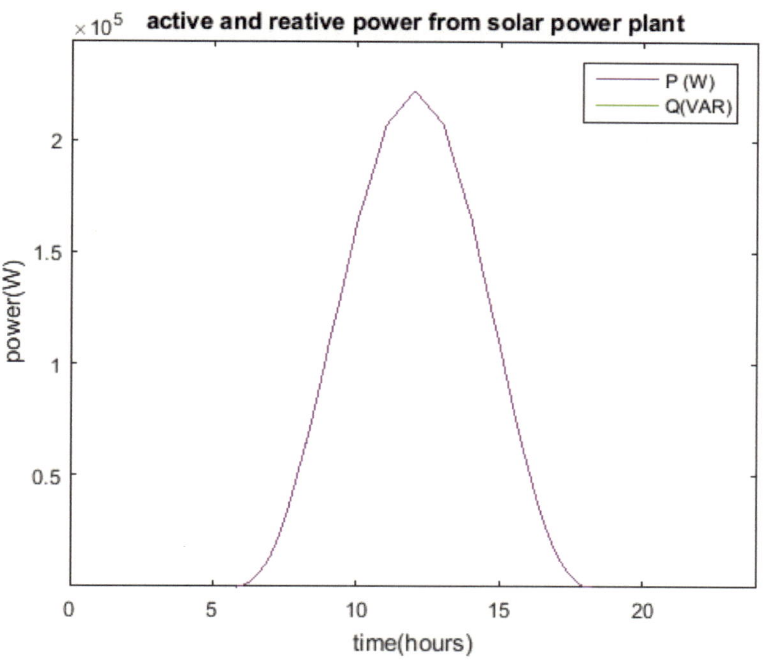

Figure 5.
Power produced by solar power plant.

3.2 Microgrid operation

As indicated hereafter, the Adrar region is mainly supplied by gas turbines. But there are many isolated populations around this locality that are powered by wind and/or photovoltaic energy.

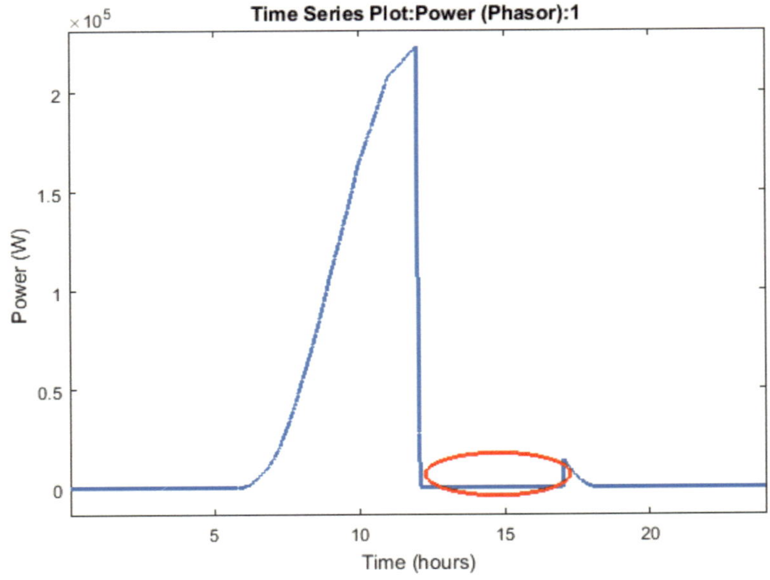

Figure 6.
Simulation of a partial shading in a sunny day from 12 am to 14 pm.

Microgrid, which integrates renewable energy sources, is connected to the main grid by relays. When the load is insufficiently powered by one of the sources, the corresponding switch will open, and other sources available will be the main providers for the load, following the scenarios below:

- Case 1: If the wind energy is sufficient to power load, wind turbine relays will be closed, and the connection to the load will be achieved.

- Case 2: Both wind and solar energies produce enough energy at the same time: the wind turbine is connected to the load, and the photovoltaic panels are used for charging the storage battery.

- Case 3: There is no half energy produced by the wind turbine, but solar radiation allows the photovoltaic panel to power the load.

- Case 4: Renewable source productions are insufficient or non-existent; the load is connected to the grid.

The aim of microgrid design is to study large-scale development, field demonstration and performance evaluation of domain names by:

- The frequency and voltage of control of different methods and technologies, under different microgrid operating modes.

- Switching between grid-connected and island mode.

- High penetration of ADR and its impact on the host network and interaction phenomena between ADRs.

It is important to note that the control strategies and dynamic behaviour of a microgrid in autonomous mode of operation may be different from those of a conventional electrical system [6, 7].

3.3 Communication used in microgrid

For reasons of reliability, large amounts of data must be exchanged as much as possible in real time between zones. This will allow the system to react quickly to any changes in network operating parameters. Microgrid control and monitoring must be ensured:

- High-speed, highly reliable and redundant communication infrastructure
- Optimal real-time communication, infrastructure
- Reliable and coordinated controls by the system before and after disruptions
- Dedicated communication infrastructure, e.g. fibre-optic cables, for monitoring underground cables
- Local flexibility market platform and algorithm

Tools based on real time synchronised network models for preventive and emergency network management systems and simulation of the behaviour of the entire electrical/energy system.

3.4 Modelling and simulation of microgrid

To improve the project's accessibility, it is required to create a microgrid model and simulate it in real time with the Opal-RT software [8].

3.4.1 Adrar main grid

Adrar grid consists of four HV lines (220 KV) and seven transformer stations.

It has six power plants with a global-produced power of 140 MW (**Figure 7**).

First, we tested the grid's operation without micro-sources in order to calculate the load rate. The result of the latter is summarised in **Table 3**.

3.4.2 Simulation of Adrar microgrid

Thanks to various metrological data (wind curve and solar irradiance in the Adrar area), we simulated the operation of the Kebertene wind farm and the Zaouiet Kounta photovoltaic plant with satisfactory accuracy, as shown in **Figure 8**.

We see that in normal climatic condition according to metrological data, we have the load entirely powered by both renewable sources during 24 h.

It is interesting for assuming that this microgrid can work in an islanded mode, during many days in the week. Diesel generator and storage can give the autonomy for the entire system.

3.4.1 Control of supply sources

We propose for supply sources control an electronic power switch (Mosfet or IGBT), controlled by a signal from an algebraic loop.

If the voltage value from decentralised sources is equal to the voltage reference, a positive signal is sent to the switch to be on.

If this is not the case, the main power system will supply the loads, as shown in **Figure 9**.

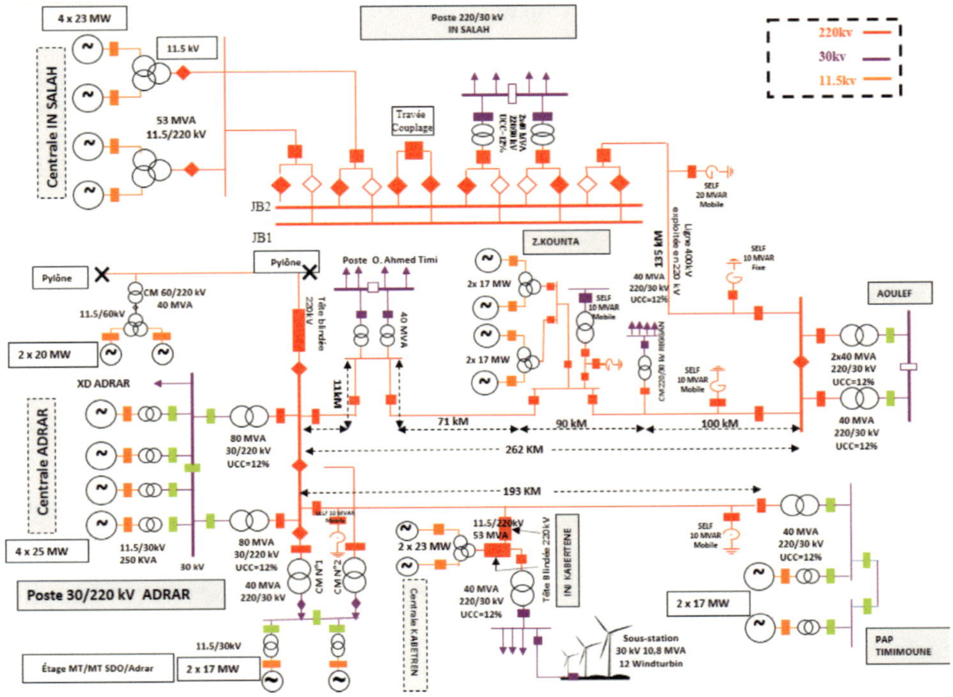

Figure 7.
Adrar grid.

Generation report		
Subnetwork 1	**Adrar network power flow**	
	P (MW)	**Q (Mvar)**
B11.5	6.12330567	167.444605
Bus 30 KV	2.08064306	38.8766705
Total	8.20394873	206.321276

Table 3.
Generation report.

3.4.2 Simulation results

The control of this system is able to verify the following three steps under climatic conditions such as:

- The wind turbine is under optimal operation conditions, so it is connected to the load.
- If both wind and photovoltaic energy installations produce enough energy, wind turbine will be used to power the load, when photovoltaic panels ensure the charge of the storage battery.
- Wind speed is insufficient, and solar radiation produces enough energy, so the photovoltaic panels are connected to the storage battery.
- All sources are deficient, so the network is connected to the load (**Figures 10** and **11**).

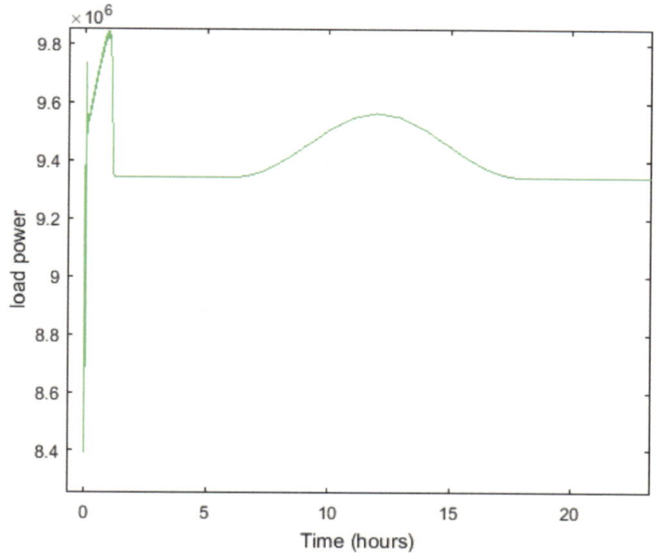

Figure 8.
Load power consumption profile.

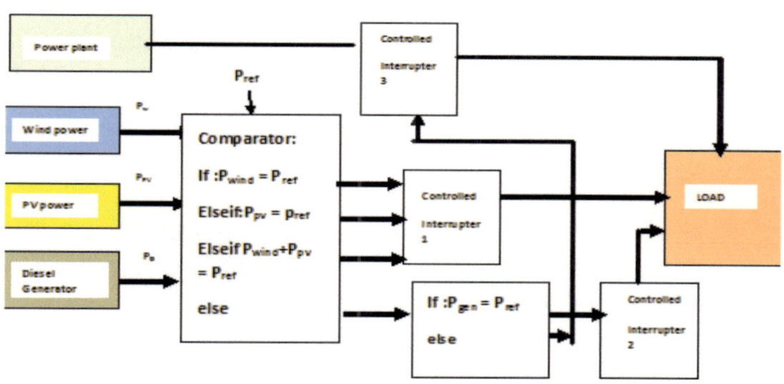

Figure 9.
Schematic microgrid control.

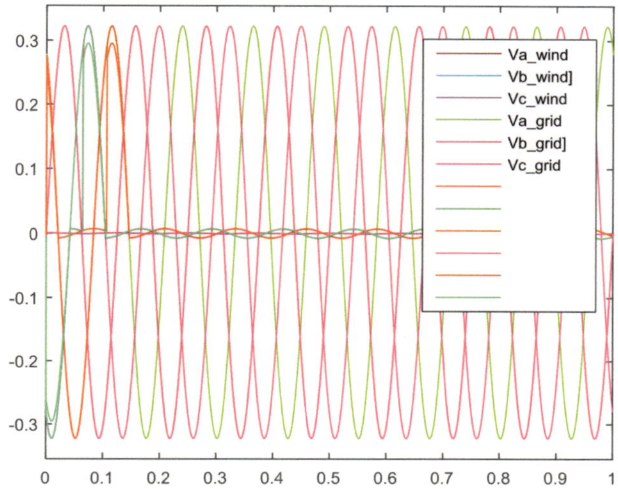

Figure 10.
Simulation of deficiency in micro-sources.

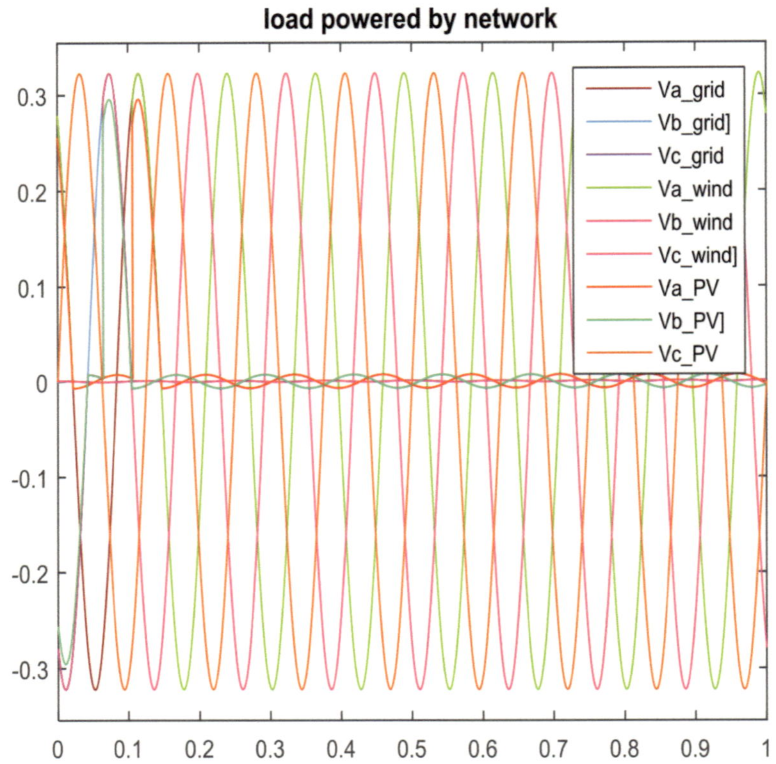

Figure 11.
Simulation of isolated microgrid.

We have simulated operating situations for Adrar microgrids, and the results confirm the expected specifications required for optimal performance.

The control and management system will provide various benefits at all voltage levels of the distribution system. For this reason, different hierarchical control strategies must be implemented at different levels of the network.

4. Real-time simulation

To improve the project's overall effectiveness, it is required to create a microgrid model and simulate it in real time using the Opal-RT software [8].

It is essential to validate this equipment before its installation on the real electricity grid. In order to accelerate the development and validation cycle of this equipment, to reduce costs and risks, the current trend is to test this equipment with a real-time digital simulator.

Therefore, the real-time simulator must reproduce as closely as possible the dynamic behaviour of the controlled electrical system [4]. The real-time simulation of the whole electrical system comes first with a modeling phase that consists of equating the system, then a design phase of an algorithmic specification (choice of the sampling period, discretization and quantification) and, finally, a real-time implementation phase [5].

4.1 Hardware architecture

The hardware system installed in our SCAMRE laboratory consists of two simulators which are connected to each other, the Wanda 4u and the OP 5600. The

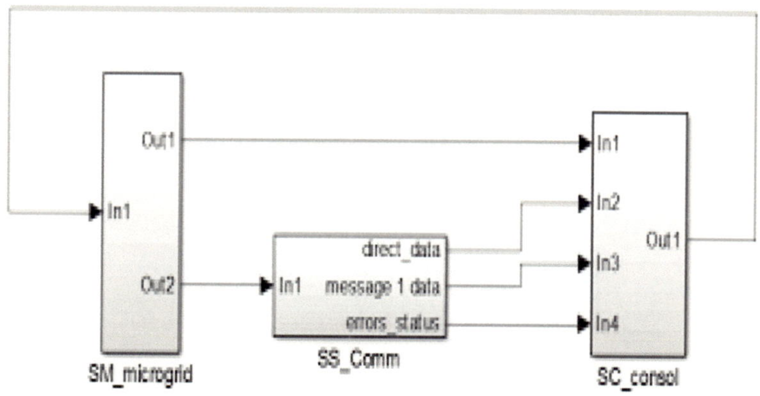

Figure 12.
Example of microgrid model with the real-time software.

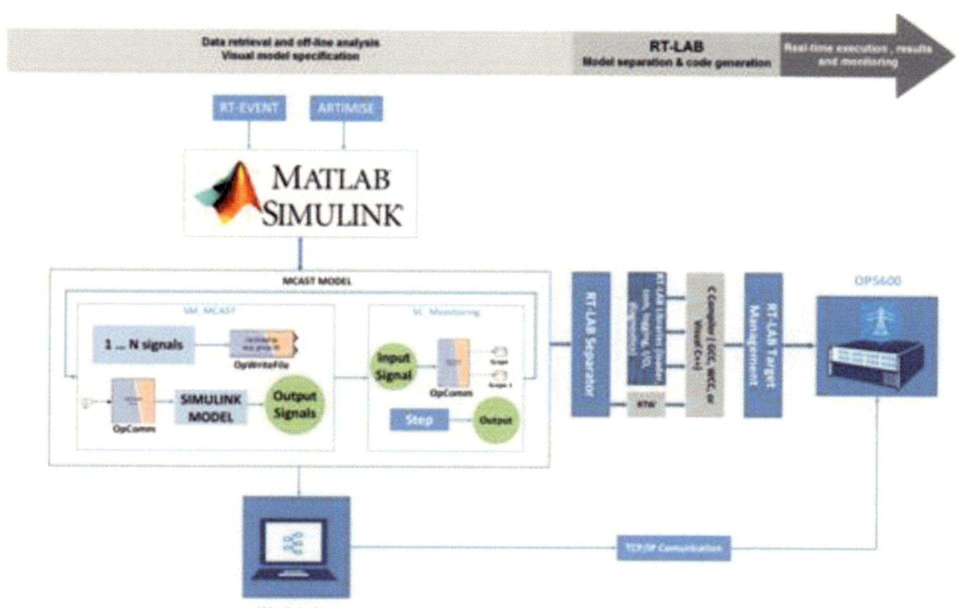

Figure 13.
System architecture in RT-LAB platform with OpComm and OpWriteFile blocks.

target has two CPU processors which have two activated cores and 16 I/O, for the first simulator, and two other CPU processors including two activated cores and 16 I/O, for the second one.

The main task of the target is to achieve a simulation of different models. The host computer will support development, editing, verification and compilation of models. Its second mission is to serve as a console or command post for control and observation during the simulation. Ethernet is used to communicate between hosts and targets. The host computer is a general PC [6, 7].

In the RT-LAB simulation platform, Artemis is a fixed time step solver designed for the electrical systems. It can improve speed simulation. The multiprocessor operating mode allows it to perform real-time simulations on the RT-LAB platform. The idea is to transform a complex system in some simple subsystems in order to perform parallel operations in a multiprocessor.

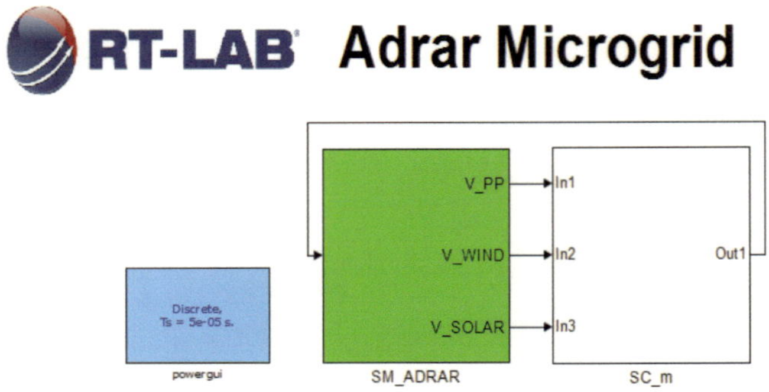

Figure 14.
Real-time compilation model.

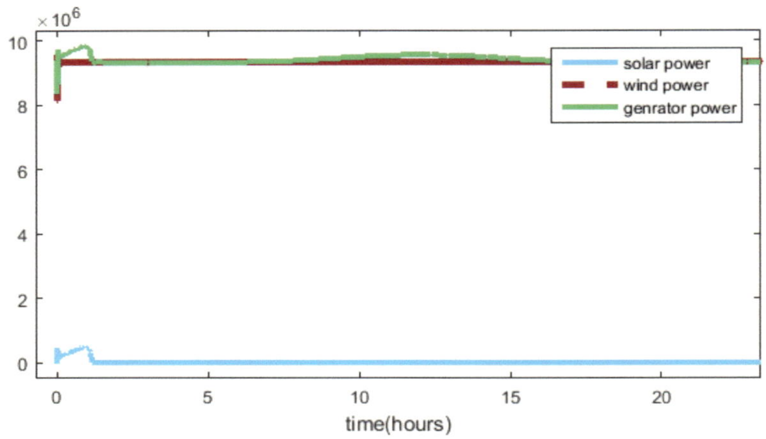

Figure 15.
Irradiance failure.

Another interesting property of RT-LAB software is the possibility to connect physical devices to the simulation system. By this way, the simulation becomes closer to reality, and the results obtained will be more realistic.

For this reason, the entire model must be reorganised mainly into three subsystems, namely, the master, slave and console subsystems, as shown in **Figure 12**.

The microgrid system is modelled first in Matlab/Simulink/SimPowerSystems software, and then it will be compiled with the e-MEGAsim simulation of the RT-LAB platform [2, 6, 7], which improves the simulation of increasingly large systems with real-time performance on multiple CPUs (**Figures 13** and **14**).

One of the main differences that improves with real-time simulation is that we can observe how our system works during 24 h. In addition, it is possible to simulate many real scenarios that affect the normal distribution of microgrids such as the disappearance of the sun for 1 h during the day, point overloads, court circuits.

Figure 15 shows the case of a sudden failure of the irradiance and the ability of the microgrid to switch to wind power to power the load.

5. Conclusion and perspectives

In this work we have designed and simulated a microgrid in real-time situation to propose the best scenario in terms of renewable sources to be installed and ability of the microgrid to operate in island mode or not.

The results obtained confirm that Saharan climate (sunny and windy) open big perspectives to integrate many autonomous microgrids in several remote areas without the need to connect them to the main grid, if there is at least a diesel generator in addition to renewable for particular situations.

The application of micro-sources can obviously regulate the consumption of distribution and transmission installations [9].

The particular landscape of southern Algeria is relevant to implement a diversity of energy sources in microgrids in order to optimise their operation and facilitate their control.

Applications of autonomous microgrids for remote areas are mainly realised for the electrification of electrically nonintegrated areas, such as, islands, or the Algerian Sahara.

A few years ago, some communities in the Sahara were supplied almost exclusively by diesel generators. In addition to reducing fuel costs, the main objective of stand-alone microgrid applications is to study and develop a field experience with the planning and operation of stand-alone distribution networks [10–12].

This work is the first conception of a microgrid in Algerian Sahara area.

It includes diesel generators, wind and solar energy.

The simulation performed in real time for this model provided us real data to improve local reliability. Gas emissions will be reduced, and power quality will be improved by supporting voltage, reducing voltage dips and potentially reducing energy supply costs.

On the basis of the promising findings presented in this paper, the work on the remaining issues is continuing and will be presented in future papers.

Acknowledgements

This work is a part of SCAMRE laboratory (Polytechnic School of Oran ENPO) research project on microgrids operation.

Author details

Mounir Khiat*, Sid Ahmed Khiat, Mohamed Mankour and Leila Ghomri
Abdelhamid Ibn Badis University Mostaganem, ENPO Oran, Mostaganem, Algeria

*Address all correspondence to: leila.ghomri@univ-mosta.dz

References

[1] Boudghene Stambouli A, Khiat Z, Flazi S, Kitamura Y. A review on the renewable energy development in Algeria: Current perspective, energy scenario and sustainability issues. Renewable and Sustainable Energy Reviews. 2012;**16**:4445-4460

[2] Zacharia L, Kyriakou A, Hadjidemetriou L, Kyriakides E, Panayiotou C, Azzopardi B, Martensen N, et al. Islanding and resynchronization procedure of a University Campus Microgrid. In: 2018 International Conference on Smart Energy Systems and Technologies (SEST) Sevilla. 2018. pp. 1-6

[3] Boulakhras C. Nouveau programme national de développement des énergies renouvelables (2015-2025). In: National Meeting with MICLAT 02 June 2018, Algeria. 2015. pp. 1-6

[4] Algerien P, Renouvelables E. Evolution du Réseau de transport de l' électricité en Algérie. 2014. pp. 1-2. Available from: www.cder.dz

[5] Khiat M, Ghomri L. Real time of HVDC and VSC-HVDC models: Application to Algerian-Spanich power system interconnection. In: ICREPQ'16. Madrid, Spain: Elsevier; 2016

[6] Himri Y, Boudghene Stambouli B, Draoui B. Prospect of wind farm development in Algeria. Desalination. 2009;**239**:130-138

[7] Mounir K, Kamel B, Ilyes K. Modeling and real time simulation of wind power systems using Rt-Lab platform. In: Opal-Rt Conference, Paris. 2017

[8] Su HC, Chang GW, Huang HM, Jen KK, Chun GC, Wu GZ. Analysis of Wind Generation System by Real-Time Simulation. Taiwan: National Chung Cheng University Chia-Yi; 2012

[9] User's Guide. Real-Time Workshop 7. MathWorks, Inc.; 2010. Available from: www.mathworks.com

[10] Zhang P, Li F, Bhatt N. Next generation monitoring, analysis, and control for the future smart control center. In: IEEE Transactions on Smart Grid; Vol. 1; No. 2. 2010; pp. 186-192

[11] Huang Y, Mao S, Nelms RM. Adaptive electricity scheduling in microgrids. In: Proceedings of IEEE Global Communications Conference; April 2013; pp. 19

[12] http://portail.cder.dz/spip.php?article3811

Chapter 3

Assessment of the Main Requirements and Characteristics Related to the Implementation of a Residential DC Microgrid

Lucia-Andreea El-Leathey

Abstract

A generic DC microgrid consists of a number of electric generators with static converters as interface modules, electric loads (to be connected either at DC or AC with inverter modules), as well as connection (by transformer and conversion modules) to the electric distribution network. The chapter envisages the state of the art on DC electric power distribution systems by tapping both high- and low-voltage direct current technologies and leading to the current development prospects. Moreover, a study on the existing standards applicable to DC distribution systems is achieved. The chapter leads to the establishment of the main technical requirements and characteristics suitable to the implementation of a residential DC microgrid. Also, electrical diagrams of the foreseen solutions and users' recommendations and challenges are suggested by the paper.

Keywords: microgrids, residential users, HVDC, LVDC, wiring diagram

1. Introduction

The direct current (DC) electric distribution system mainly consists of converters and DC links, as suggested by [1]. The DC distribution systems can be classified as high-voltage direct current (HVDC) or low-voltage direct current (LVDC) for which the DC-AC conversion is located near the end users.

Currently, high-voltage direct current systems are widely used for offshore and submarine electric power transport, [2], in order to interconnect non-synchronized AC grids, thus providing efficient and stable transport and control capacity. HVDC also represents the suitable technology for long-distance energy transport and large amounts of power with reduced power losses [3]. Moreover, HVDC systems stand as a key concept for overcoming the existing barriers concerning the energy generation from renewable sources (wind, solar, or hydropower), since these generating units are rarely located near urban areas or domestic energy consumption points.

The selection of a HVDC topology dedicated to the transport of large amounts of electric power over long distances in a specific case is due mostly to the following aspects:

a. Power control: HVDC is necessary from the technical point of view in order to ensure intelligent control.

b. Reduced total investments: the use of HVDC for the interconnection of two nodes within an electric power system is in many cases the more economical alternative compared to high-voltage alternative current (HVAC) systems as shown in **Figure 1**.

c. Environmental protection

Thus, HVDC systems allow the safe and stable interconnection of AC electrical grids operating at different frequency values, [5], which are otherwise incompatible. In addition, HVDC provides instant and accurate power flow control.

As far as it goes, LVDC microgrid concept has gained the scientific community attention in recent years. A direct current distribution microgrid represents a practical solution to efficiency problems of existing AC electrical grid [6]. While various papers have shown that DC microgrids can play an effective role in solving some operational

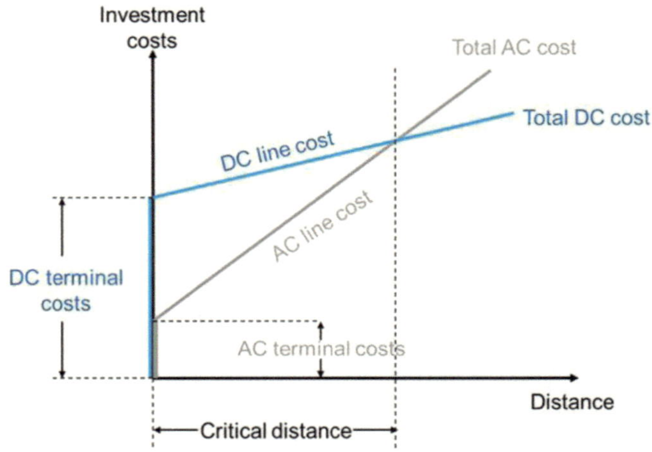

Figure 1.
Cost versus transmission distance for HVDC and HVAC systems [4].

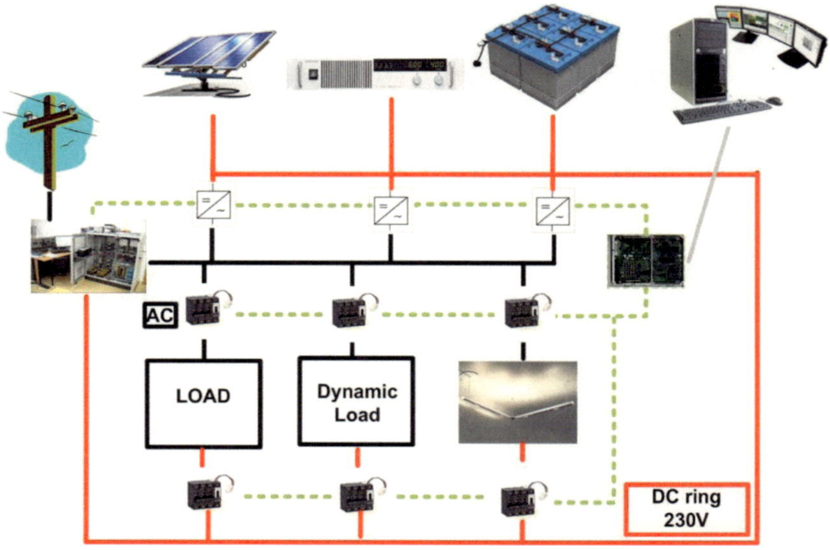

Figure 2.
The key diagram for a mixed microgrid with separate AC and DC rings.

issues on the main grid, [7, 8], one paper in particular envisages that a DC microgrid can be used for voltage support, by providing the option of injecting reactive power as an ancillary service [8, 9]. Consequently, the most important aspects regard the integration of renewable energy sources and energy storage systems dedicated to individual end users, in order to improve the quality, reliability, and energy efficiency, respectively [10]. Moreover, the increasing number of DC household appliances has brought to the attention both AC and DC distribution systems, as represented in **Figure 2**.

Thus, the chapter analyzes researches on DC distribution systems as well as their potential for residential applications. General advantages of LVDC distribution systems are envisaged, highlighting various power supply architectures and topologies. Also, there are presented demonstration facilities in which LVDC distribution systems have been implemented.

2. Analysis of existing standards applicable to DC operating power supplies

The following standards are used by the European Telecommunications Standards Institute (ETSI) for the coordination of Telekom stations equipment. These normative documents are required to be applied for DC operation:

- *IEC/EN 60947-2:2009 Low-voltage switchgear and control gear—Part 2: Circuit-breakers*

This standard applies to automated circuit breakers, of which main contacts are intended to be connected to circuits, the rated voltage of which does not exceed 1000 V AC or 1500 V DC; it also contains additional requirements for integrally fused circuit breakers [11].

- SR EN 60947-3:2009 *Low-voltage switchgear and control gear—Part 3: Switches, disconnectors, witch-disconnectors and fuse-combination units*

This standard applies to switches, disconnectors, switch-disconnectors, and fuse-combination units to be used in distribution circuits and motor circuits of which the rated voltage does not exceed 1000 V AC or 1500 V DC. The manufacturer will specify the type, ratings, and characteristics according to the relevant standard of any incorporated fuses [11].

IEC 60947-2 standardizes miniature circuit breakers (MCB) used by the industry. The aforementioned standard creates a frame for power distribution with voltage values of up to 1000 V AC and 1500 V DC for all rated current ranges starting from 0.5 and reaching 6300 A. There are three types of circuit breakers in this class:

- Air circuit breakers (ACBs)
- Molded case circuit breakers (MCCBs)
- Miniature circuit breakers (MCBs)

An alternative standard for automatic circuit breakers is represented by IEC 60898-1, which refers to the miniature circuit breakers used within the low-voltage AC grids and included in electrical panels from households, shops, or office buildings. **Table 1** synthetizes the major differences between the two standards which apply to MCBs.

Another difference is to be observed when comparing the trigger curves. If the IEC 60898-1 standard clearly describes the B, C, and D curves as depending on the rated

MCB's characteristics	IEC 60898-1	IEC60947-2
Regulated for:	**Residential domain**	**Industrial domain**
Rated current, I_n	6–125 A	0.5–160 A
Maximum current, I_{cn}	25 kA	30 kA
Rated voltage, U_e	400 V	440, 500, 690 V
Impulse voltage, U_{imp}	4 kV	6 or 8 kV
Degree of protection	2	3
Trigger curves	B, C, D	B, C, D, K, Z, MA
Operation mode	AC	AC or DC
Maximum ambient temperature	30°C	50°C
Electrical auxiliaries	No	Monitoring and control

Table 1.
Comparison between the characteristics of miniature circuit breakers (MCBs) regulated according to IEC [12].

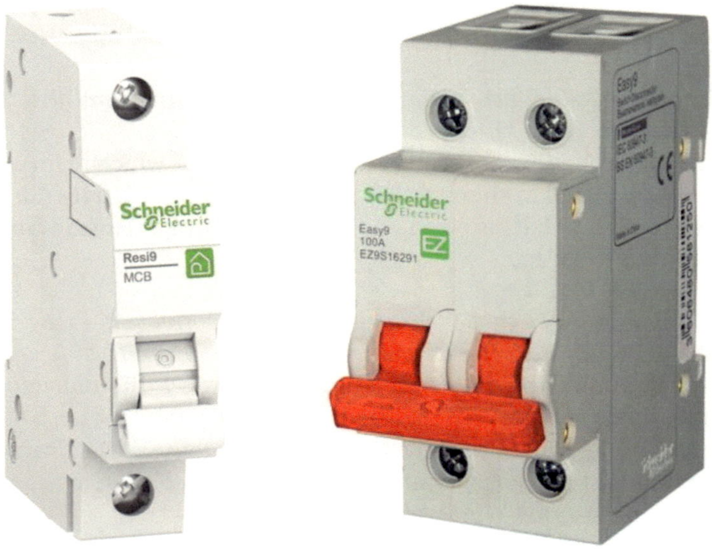

Figure 3.
Examples of Resi9 circuit breakers for residential applications and Eazy9 for industrial use manufactured by Schneider [13].

current, IEC 60947-2 shows that instantaneous triggering may be adjustable in accordance to the user's necessities or predefined by the manufacturer with a precision of 20%. This is the reason why many manufacturers have added K, Z, and MA curves (**Figure 3**).

In conclusion, the use of MCBs which are certified according to both standards and are suitable for residential as well as industrial use is preferable.

- *IEC/EN 60269-1 Low-voltage fuses—Part 1: General requirements*

This standard is applicable to fuses incorporating enclosed current-limiting fuse links with rated breaking capacities of not less than 6 kA, intended for protecting power-frequency AC circuits of nominal voltages not exceeding 1000 V or DC circuits of nominal voltages not exceeding 1500 V.

The standard has been updated and released again as SR EN 60269-1:2008/A1:2010 and SR EN 60269-1:2008/A2:2015.

- *IEC/EN 61000-4-29 Electromagnetic compatibility (EMC)—Part 4–29: Testing and measurement techniques—Voltage dips, short interruptions and voltage variations on DC input power port immunity tests*

- *IEC/EN 61000-4-5 Electromagnetic compatibility (EMC)—Part 4–5: Testing and measurement techniques—Surge immunity test*

As mentioned by [14], in the United States, restrictions regarding electrical systems are established by the requirements of NEC (NFPA-70), Section 210.6, for branch-circuit voltage limitations. Thus, this code from the US National Electrical Code imposes the voltage in residential and similar occupancies to 120 V, while only in the case of specific loads, the limits are raised to 277 V among conductors.

Also, at the moment, an IEEE standard for DC microgrids for rural and remote electricity access applications is under development, targeting sustainable DC off-grid and remote power and relying firstly on user's safety.

3. Assessment of consumption requirements and preliminary features for the development and implementation of residential DC microgrids

3.1 Methods for adapting the electrical consumers to DC grid parameters

In order to establish the proper methods regarding the consumers' adaptability to a DC grid, it is necessary to identify the ones which are to be used and to analyze the possibilities of modifying the included power supply. The classification of electrical appliances that are usually found in a household is therefore highlighted. The main categories of electric consumers can be defined as follows:

- Heating appliances with resistive loads only (hotplates, ovens, radiators, heat exchangers) can easily be supplied from the DC grid if the same voltage level and electric power requirements are ensured as from the single-phase AC grid. If the DC voltage is different from the end users' parameters, buck (step-down) or boost (step-up) DC/DC converters can be used.

- Household appliances with inductive loads (fridge, pumps, vacuum cleaner, fans, etc.) using mainly asynchronous AC motors require DC/AC inverters in order to restore the pure or modified sine wave which drives the electric motors.

- Equipment with multiple (resistive, inductive) loads: washing machines and HVAC equipment (air conditioning, ventilation/heating) which operate by using a DC/AC inverter in order to adapt the required operating parameters. These household appliances also include elements that can be DC power supplied, but the required separation of the DC and AC paths is difficult and not justified.

- Low-power equipment using DC in the voltage range of 5–48 V. These can operate with step-down DC/DC converters. A separate 48–50 V power supply line can be used for a low installed power supply for the following end users: LED, laptop/desktop PCs, telecommunication equipment (router, mobile phone), LED TV, monitors, printers, etc.

A step-down converter with 48 V input and 5–9–12–20–24 V output operates with very good conversion efficiency. Due to the fact that it does not process significant power amounts, it has an affordable price.

Currently, the home appliance industry is mainly focused on AC power supplied products. Still, there are an increasing number of DC power devices that use switching mode power supplies (SMPS) for AC conversion and voltage level adjustment. These devices can be modified by eliminating the rectifying and power factor correction modules.

SMPS are electronic power sources which include switching regulators for the efficient energy conversion. SMPS use a transistor (or a group of pass transistors) that continuously switch between low-dissipation, full-on, and full-off states in order to remain as little as possible in high-dissipation transitions, thus minimizing the wasted energy. Ideally, switching mode power supplies do not dissipate any power. Voltage regulation is achieved by varying the time ratio between saturation and blocking. High power conversion efficiency represents an important advantage of a switching power supply. Also, SMPS can have significantly reduced dimensions and be lighter than a linear power supply due to the size and the weight of the included transformer. Switching regulators are used as replacements for the linear regulators when higher efficiency or more reduced size and weight are required. These are, however, much more complicated; if not suppressed, current can cause electrical noise problems during switching, while the simple models may have a low power factor. Ideal switching elements (e.g., transistors operating outside their active mode) have no resistance when "open" and do not carry any electrical current when "closed." Therefore, converters are able to theoretically operate with 100% efficiency (e.g., all input current is delivered to the load, and no current is wasted as dissipated heat).

The output of the switching source is adjusted by using the fill factor control; the transistors are switched in fully closed or open stages, so that the resistance losses between input and load are limited. The only amount of generated heat results from the non-ideal characteristics of the used components and from the residual currents related to the control circuits.

The losses due to transistors' switching (especially in the short part of each cycle when the device is partially activated), the switching transistors' resistance, the series resistance of both the inductor and capacitors, and the inductor's iron losses as well as the voltage drop on the rectifier diodes lead to a specific efficiency of about 60–70%. However, the optimization of the SMPS design (choosing the optimal switching frequency, avoiding inductor saturation, and active rectifying) will provide the minimization of the energy losses. Thus, an optimal switching source configuration will be characterized by 95% efficiency.

The efficiency of the DC/DC converters is comparable to that of the switching sources if considering that the operating principle is similar after the point of power factor rectifying and correction. Most currently manufactured SMPS also include power factor compensation circuits in order to reduce grid losses and disturbances and to comply with international regulations.

The unity power factor represents the objective of any power generating company, because otherwise, a higher current value has to be provided to the end users for a certain power demand. In this respect, the manufacturer sustains higher line losses. In the case of an industrial power plant, a penalty is charged if the power factor is way different from 1 (under the neutral power factor of 0.92). Mainly, motors' windings act as inductors within the public distribution grid. Opposite effect capacitors which are compensating the motors' inductive windings can be used.

SMPS do not operate as reactive loads like the electric motors but instead represent nonlinear loads for the power supply grid. Sources without power factor correction (PFC) absorb high current pulses or spikes from the AC grid (which provides sinusoidal voltage) due to the low conduction angle in the input stage that carries out the rectifying. If left uncompensated, a switching source power factor (PF) will generally be equal to 0.65 or even lower. PF can be compensated by using power factor correction circuits. These circuits smooth current pulses, improve PF, and reduce the possibility of the AC circuit breaker safety devices to act prematurely.

There are two basic types of PFCs: passive and active. Passive PFC circuits are less expensive and usually can compensate the power factor at around 0.85. The PFC active circuits are the most used ones and are even included by the power supply source, thus increasing PF over 0.98. A close to 1 PF indicates good power supply performance.

Due to the high increase of household appliances that include power supplies which add up to existing consumers, since 2001 the European Union (EU) has set harmonic currents' limits that can occur within the AC power grid and are caused by SMPS.

The most important regulation is EN61000-3-2 which relates to SMPS with input power of over 75 W while absorbing up to 16 A electric current. Severe limits regarding up to the 39th harmonic currents, measured at the power supply input, are set. For example, EU has established a 50 Hz value for the frequency of the first harmonic. The third harmonic is equal to 150 Hz, while the 39th harmonic equal to 1950 Hz. These unwanted harmonic currents have a direct connection to the SMPS power factor. PFC significantly reduces the AC harmonics, leaving mainly the "fundamental," which is in phase to the waveform. Power supplies that meet the EN61000-3-2 standard are normally characterized by a power factor higher than 0.97.

PFC increases the power supply capacity, thus determining the amount of useful energy which SMPS will use from the AC grid and then deliver it to a load. The relation showing the abovementioned is shown in Eq. (1):

$$P_{out} = V_{RMS} \cdot I_{RMS} \cdot PF \cdot Efficiency \quad (1)$$

Many technologies and topologies can be and were designed for PFC. When dealing with low installed energy (even up to 200 W), various passive PFC techniques have been used in order to increase the conduction angle of the electric current waveform [15].

It can therefore be concluded that if SMPS include the power factor correction module, the conversion efficiency will be similar to the case of DC/DC converters. In the absence of power factor correction circuits, the DC power supply will contribute to reducing losses in the power transformation chain.

Moreover, if the foreseen household DC grid is characterized by a nominal voltage that coincides with the rectified voltage of an equipment switching source, then the appliance is ought to be directly connected to the DC power grid without any further changes.

Recent developments in the field of electric drives have allowed the large-scale use of Inverter technology for most household appliances using significant power motors. The main purpose of this technology is to increase the energy efficiency by varying the motor speed which can also operate on partial loads. The Inverter technology represents the most recent technological evolution regarding the compressors' electric motors. The inverter is used in order to continuously adjust the temperature by controlling the speed of the compressor motor. The DC inverter units dispose of a variable frequency drive, comprising an adjustable electric inverter which controls the speed of the electric motor and of the compressor, respectively. The unit converts the AC current input to DC, and then by modulating it through an inverter, it produces the foreseen current frequency. A microcontroller will acquire each value of the ambient air temperature and therefore adjust the compressor speed accordingly. The inverter air conditioning systems operate more efficiently than conventional ones, providing extended lifetime of their components

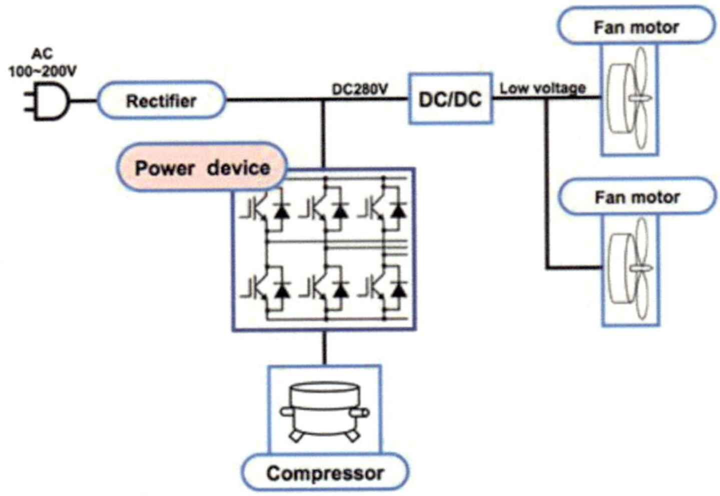

Figure 4.
Inverter technology freezer developed by Mitsubishi [16].

and not introducing disturbances to the main grid. An example of a freezer Inverter technology implemented by Mitsubishi is shown within **Figure 4**.

As shown in **Figure 4**, the system includes an AC/DC rectifying module, subsequently used in order to reshape the sinusoidal waveform along with the variable frequency required by the compressor and fan motors. It can thus be directly supplied with adequate DC voltage without any additional problems.

In the case of DC-powered equipment with low voltage levels, it has been shown previously that DC/DC converters can be used successfully.

DC/DC converters represent power supplies that convert electrical power with an unstable DC input voltage into a stabilized DC output voltage, to different values, lower value (step-down), higher value (step-up), and equal (stable level), or to inverse polarity comparing the input voltage (invert).

The more complex converters are based on microcontrollers in order to ensure high efficiency and as low as possible size, disturbances, and thermal dissipation losses. DC/DC converters are generally used in order to isolate electrical noise, for galvanic isolation, to voltage level conversion, and to provide a stable voltage level for voltage-sensitive equipment and various battery voltage values which supply portable equipment. Power density, efficiency, and reliability represent the basic characteristics which are considered for the price/performance ratio evaluation of a DC/DC converter. DC/DC converters are widely used either for fixed equipment power supplies (supplied from the AC grid) or for portable (battery powered) and IT equipment (where various voltage values are required for CPU, RAM, memory drives, and interfaces) [17].

DC/DC converters are basically SMPS with the following advantages:

- Very high efficiency comparing to the case of linear sources (typically 75–90%).

- Reduced energy transfer loss since all the components are smaller and require simple thermal management.

- The energy stored by a coil from a switching regulator can be supplied with higher voltage than the input voltage (boost) or even negative (inverted);

when using a transformer, galvanic isolation can be provided (minimum 1000 V DC), which cannot be achieved by a linear source.

The basic operating principle of a DC/DC converter regards the command of a high frequency switching element (at least 100 kHz) by variably controlling the on-time/off-time ratio ("duty ratio") in order to keep the output voltage at a certain value. Usually, the voltage is constantly controlled through the negative feedback of the output voltage. Some switching sources also solve the problem of electrical noise, with specialized controllers embedded in integrated circuits.

It is worth mentioning that about 80% of electromagnetic compatibility (EMC) problems are due to both power and I/O cables which produce an unintended "antenna structure." This structure can emit the electromagnetic energy generated by product-embedded electronic components and also receive the electromagnetic energy from the product's exterior. EMC regards electromagnetic interference (EMI) which stands as the amount of emitted energy, whether intentionally or otherwise, by electronic equipment that cause performance degradation on nearby equipment. Also, EMC addresses electromagnetic susceptibility (EMS) and the lack of immunity to internal or external interference, respectively. Emissions of radiated or conducted disturbances (within the AC grid) by IT equipment are covered by EN 55022 standard.

EN 55024 and EN 61000-4-2, 3, 4, 5, 6, and 8 standards regulate immunity to electrostatic discharges (ESD), intentional radio emissions, switching noise or electrical transitory regimes, lightning, 50–60 Hz variable magnetic fields, and power fluctuations in the AC grid [17].

The maximum efficiency of the operating switching sources is associated to a well-designed load when the equipment is power supplied similarly to the nominal regime parameters, yet considering a certain reserve of power. For example, a source works seamlessly with a load of only 10%, but energy conversion losses are higher than in the case of 80–90% load.

Therefore, manufacturers have developed switching sources suitable to a wide range of products provided with different input/output voltage values while characterized by efficiency between 70 and 96% and a few watts up to several thousand watts power.

Also, there are available various sources with a wide range of input and output voltage control according to the consumer's parameters. Thus, the same source is able to supply equipment that work at 5, 9, 12, 20, 24, and 48 V. The voltage can be adjusted before coupling, whether it is the case of a single output related to an only consumer or the case of a single source ensuring different simultaneous voltage values through dedicated terminals.

A solution for integrating these DC/DC converters in order to supply office equipment is shown in **Figure 5**.

From 24 V DC, the voltage can be reduced to the required voltage in order to power each device. A DC/DC converter can be mounted in a distribution panel for adapting the voltage from 380 to 24 V DC. Then, by using a low-power converter, the power can be transmitted to consumers through a device that integrates multiple outlets. Such a device is shown in **Figure 6**.

In order to power up electrical consumers with different voltage values such as laptops, monitors, and mobile phones that require voltage below 24 V AC, a buck converter can be integrated into a compact unit that incorporates multiple outlets. The coupling terminals for each voltage are integrated into the socket that can supply up to 100 W.

When considering only AC operating equipment, DC/AC converters (inverters) must be used. This solution is more complicated in terms of power electronics and thus more expensive. Inverters are provided with DC voltage (or current) source as input that converts it into alternating voltage (or current) for output, which can

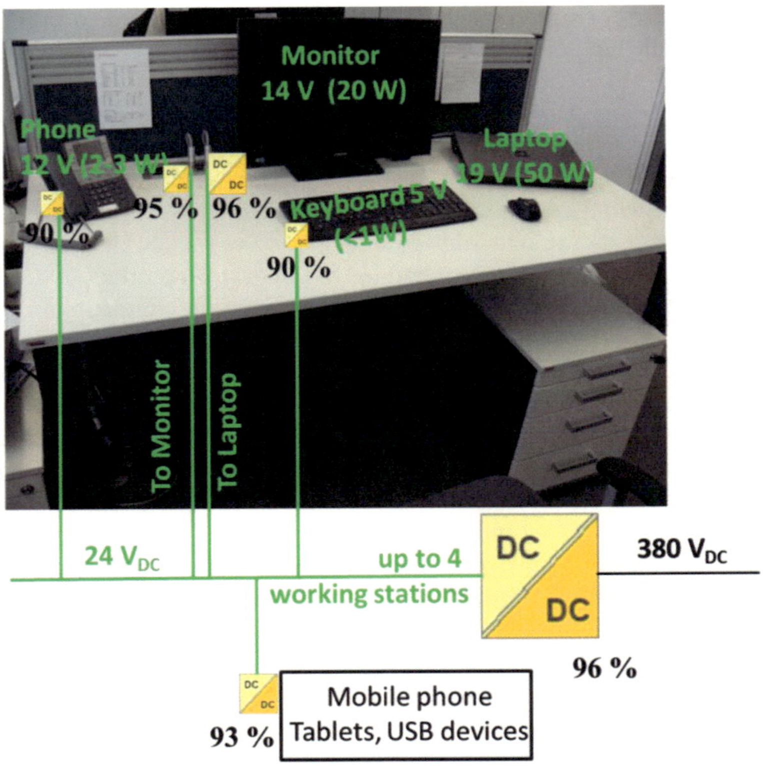

Figure 5.
Power supply infrastructure for office equipment [18].

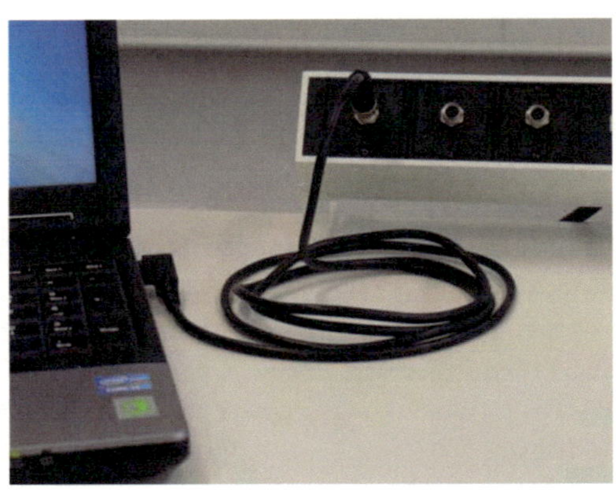

Figure 6.
Power supply device for low-power equipment [19].

have adjustable frequency and/or voltage. Usually, inverters are used to drive AC motors with adjustable rotational speed but are also applicable to other domains, for example, the case of uninterruptible power supplies (UPS). It is therefore necessary to convert the voltage value of about 300 V DC to 220–240 V AC. The on the market wide available solutions usually convert low voltages (12/24/48 V DC) due to the fact that they are required in backup systems which use battery storage for the electric power generated by photovoltaic panels or wind/water turbines. The fact that there is no market for 300 V DC/240 V AC poses a challenge and is due to this system's high price which needs to be produced on special orders. Standard inverters that accept low voltage inputs may be used, but the considered electric grid must support higher current values for the same transferred power. It can also be used along with step-down converters, but the solution is costly and inefficient increasing the losses on the power transformation chain.

All of the aforementioned aspects show that switching home appliances to DC power configurations can be achieved without any significant problems or at significant costs.

The transition is easier when using new equipment that mainly embeds switching power supplies or *Inverter* technology to drive washing machine motors or compressors for refrigerators and air conditioning systems.

Older generation equipment that still uses transformers or AC motors requires more significant changes of the power sources or the use of DC/AC converters.

These aspects do not stand as obstacles to the development of DC grid when considering that the aging of old appliances will gradually eliminate them and lead to their replacement with newer technology that is easily adaptable to the DC grid.

3.2 Considerations regarding the assessment of a typical household required power

According to Siemens, buildings account on about 40% of global energy consumption and 21% of greenhouse gas emissions, respectively. Consequently, buildings represent the key to reduce energy consumption and support sustainable urban development. The use of modern technology in intelligent buildings can reduce emissions to 40% without affecting the comfort of residents.

Thus, the intelligent house concept is evolving in response to technological progress regarding distributed energy sources as well as information and telecommunication technology, so that through management systems, consumers can contribute to more efficient use of electric power. The use of smart home energy management systems will allow the end users to efficiently use low-cost electric or thermal power [20].

According to the 2016–2030 Romania's Energy Strategy, [21], with 2050 perspectives, the electricity consumption for 2016 by its destination is shown in **Figure 7**. The used electric power in MW by activity sectors is also indicated by **Figure 8**.

According to the 2016–2030 Romania's Energy Strategy, [21], with 2050 perspectives, the share of household consumption will not change significantly. Due to technological progress, it is still possible that the energy share is related to a household to vary between certain limits depending on the appliances within.

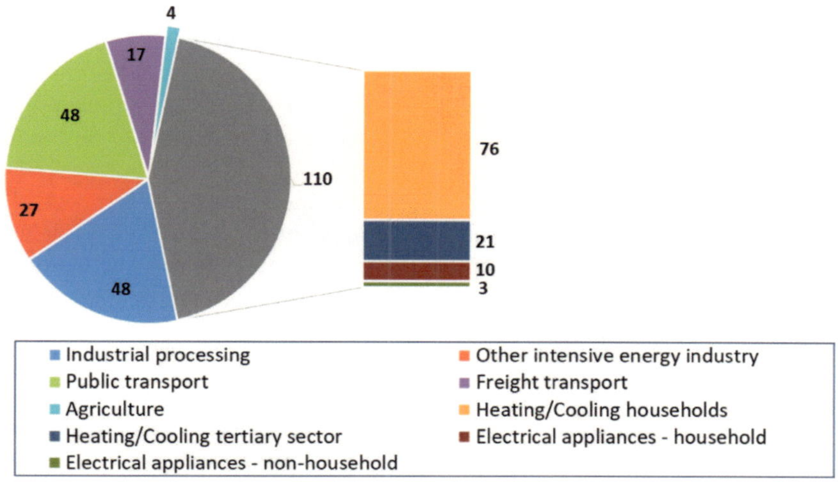

Figure 7.
Diagram regarding the final electric power consumption [MW] [21].

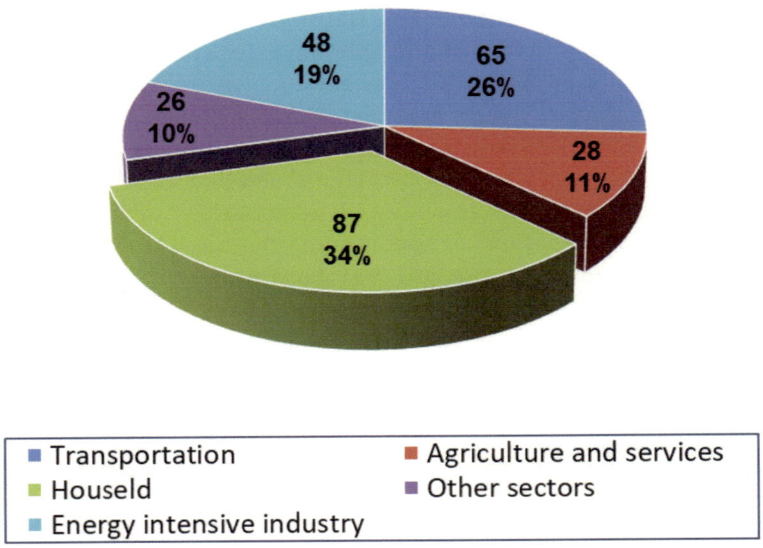

Figure 8.
Electric power [MW] by activity sectors [21].

When discussing 2016, the share of household appliance consumption is shown in **Figure 9**.

At the moment, the price cost of a new household appliance is payed off by lower operating costs (electric power, water or other forms of energy). If considering the share consumption related to household appliances beside the total electric power consumption of the residential sector as well as the life exceeded household appliances in Romania, it becomes obvious that there is a significant potential to reduce the power and water demand, and implicitly final costs, by replacing old equipment with new one, which is much more energy efficient.

Currently, when comparing to other European countries, Romania is characterized by a very low replacement rate of household appliances. In most cases, the average replacement rate exceeds the lifetime of the equipment. These aspects can also represent positive perspectives, namely, the growth potential of new energy-efficient household appliance market [21].

Hereinafter, electrical power requirement is estimated by considering various household electrical consumers and appliances generally used. The table below shows the consumers taken into account for the power estimation as well as their characteristics (**Table 2**).

Following the estimation, it is shown that 5 kW is sufficient to supply a typical household. Depending on the season, outside temperature, or time of day, actual consumption may be much lower than anticipated.

If the air conditioning is operating while at the same time an electric vehicle is charging, the consumption is higher (about 50% of the required energy). When these consumers are not connected due to the fact that there is no need for heat/cold or the electric car has enough remaining power, then the local generated energy can be stored in batteries or transferred to the public DC grid.

In order to ensure the energy needs, several power sources can be considered depending on the location's energy potential: photovoltaic panels and wind/water electric generators. A microgrid based on renewable energy sources can be technically and economically designed in order to provide 5 kW power. A 3 kW photovoltaic panel system provided with maximum power point tracking technology (MPPT) along with a 2 kW wind turbine can generate the required power. The surplus energy will be stored in batteries and used when needed. If the microgrid is designed to work on-grid and not autonomously, public AC grid access will be opted via AC/DC bidirectional converters.

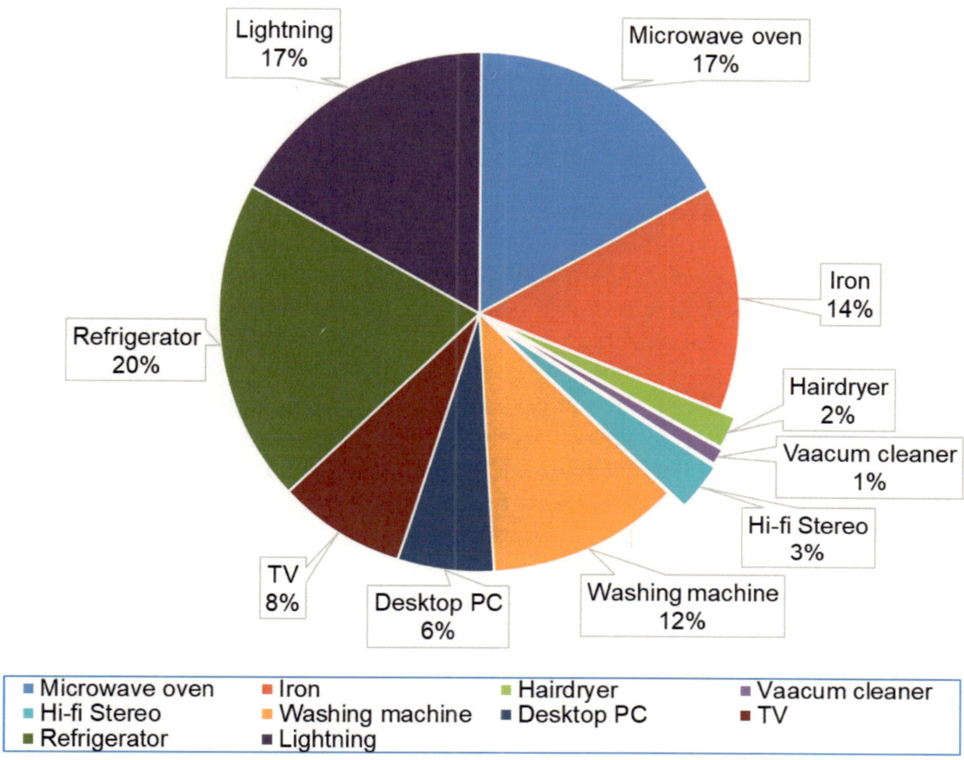

Figure 9.
The share of household appliance power consumption [21].

No.	Electric consumer	Installed power	Absorbed power	Rated voltage	Current	Using time/h	Total using time/ day	Required energy
UM		[W]	[W]	[V DC]	[A DC]	[min]	[h]	[kWh/day]
1	Heat pump	3000	750	375	8.00	15	12	9
2	Ventilation	250	83	375	0.67	20	18	1.5
3	Washing machine	2000	1000	375	5.33	30	1	1
4	Refrigerator A+	200	50	375	0.53	15	24	1.2
5	Dishes washing machine	1600	400	375	4.27	15	1	0.4
7	Lightning	300	300	48	6.25	60	4	1.2
8	TV + laptop + router	100	100	48	2.08	60	6	0.6
9	Electric vehicle charger	1500	1500	375	4.00	60	8	12
10	Power supply pump	1600	267	375	4.27	10	24	6.4
11	Total	10,550	4450	—	35.4	—	—	33.3

Table 2.
Electric consumers considered for the estimation of the necessary power.

4. Establishment of the DC microgrid layout

The previously detailed aspects demonstrate the efficiency of DC power distribution.

Although DC grids show many advantages and most home appliances are able to operate by using DC power, the development of these power distribution grids has yet to face several challenges, including:

- The necessity of using bidirectional energy conversion equipment, DC/DC and AC/DC converters, respectively.
- In-service safety and fuse protection.
- Universally accepted standard voltage regarding the operation of household appliances, as well as telecommunication equipment, electric vehicle transport, and aerospace industry.

Moreover, potential users do not yet have access to home appliances provided with DC power options. In terms of developing standards, several international organizations such as Emerge Alliance, ETSI-IEC, and IEEE are taking steps in order to establish regulations which are needed for the implementation of DC systems designed for residential applications.

According to IEC 60038 standard, distribution systems using low-voltage direct current should not exceed 1500 V. In this respect, the voltage levels' diagram which is currently used for DC power supply is shown in the following (**Figure 10**).

The transition from current distribution systems to DC grids must be gradually achieved, by ensuring parallel operation of both systems in order to ensure continuity of the power supply. DC power distribution is considered a solution

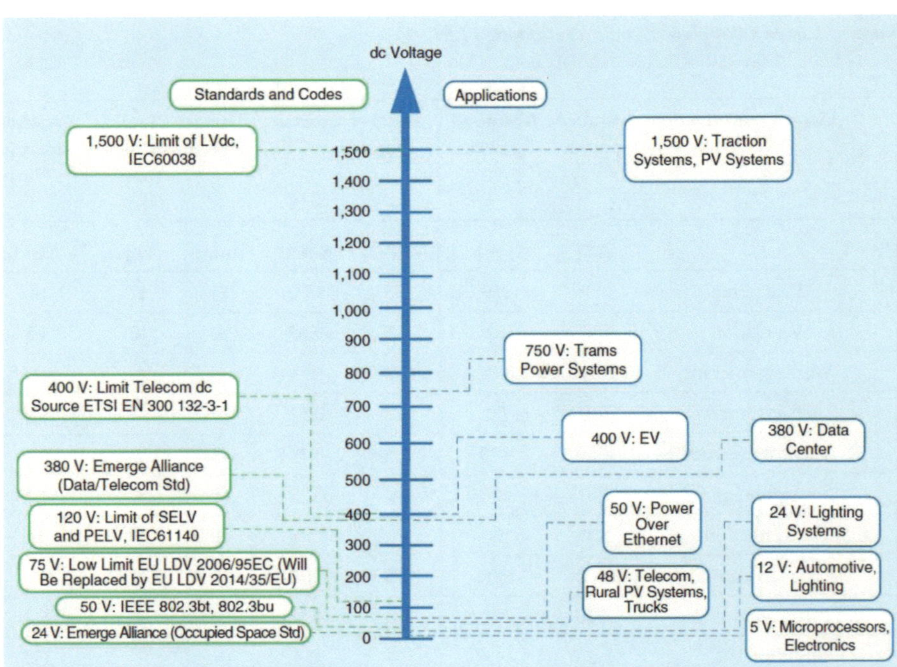

Figure 10.
Voltage levels currently adopted for DC power distribution [22].

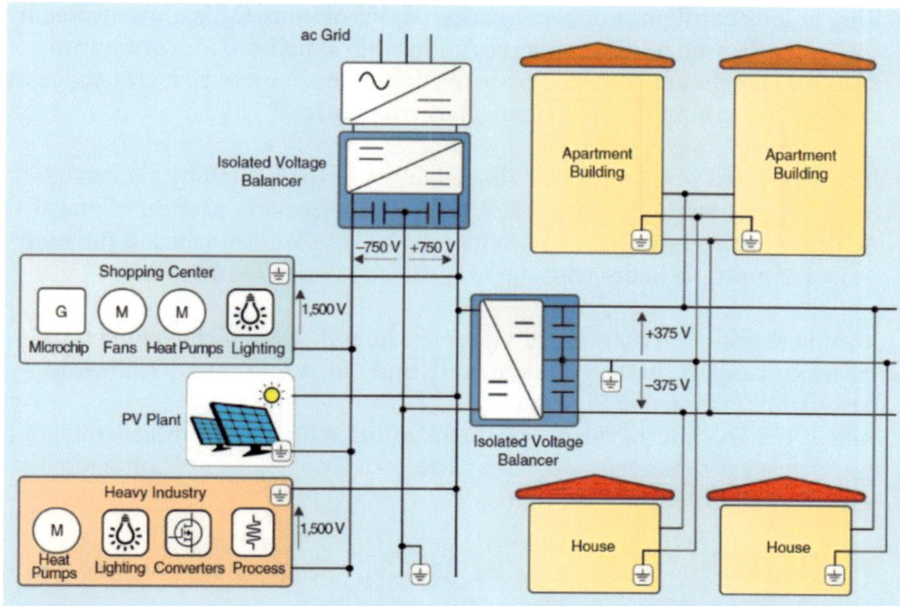

Figure 11.
DC grid provided with multiple voltage levels [22].

only if it can easily be adapted to household consumers or when it can be efficiently implemented by direct connection to renewable energy sources such as photovoltaic panels. **Figure 11** shows a DC grid structure using several voltage levels depending on the consumers' requirements.

As in the case of alternating current distribution systems, the energy flow can be transmitted through two conductors (single phase) or four conductors (three-phase). The DC grid power can also be transmitted in a similar configuration: two-wire (unipolar) and three-wire (bipolar) systems. The difference between the two grid configurations is given by the number of available voltage levels.

The need for fast integration of renewable energy sources (such as photovoltaic panels) and storage units into distribution systems has highlighted the benefits of using DC microgrids.

On the other hand, the power supply related to any distributed energy source is not time constant due to weather condition dependence. Therefore, the AC grid interface is very important in order to improve the reliability and availability of the microgrid [22].

There are several ways to connect the DC grid to an AC grid, of which there are to be mentioned:

- Radial configuration—for which the DC path starts from the AC grid through an AC/DC converter, supplying the line consumers through a single DC bus. This configuration has certain advantages such as the simple mounting and operation as well the multiple voltage level option. However, this configuration is not flexible under defective conditions. For example, a single failure in the input power end may affect all the connected consumers afterward. A radial DC grid model is shown in **Figure 12**.

Various radial DC microgrids are currently implemented and operating throughout the world. Several microgrid test beds from the United States are to be mentioned: University of Miami test bed, Florida; Sandia National Lab Test bed, Washington, DC; and UT Arlington test bed, Texas.

- Ring or loop configuration—consisting of two or more DC paths powered by converters in a single node or across multiple grid nodes. This configuration uses fast switches in order to isolate the defective circuit so that other consumers can still be supplied without any inconvenience.

- Interconnected configuration—the reliability of the DC microgrid can be further improved by providing alternative AC busses in the event of one of the AC/DC converters' failure. An example of an annular grid which is interconnected at multiple nodes with the AC grid is shown in **Figure 13**.

A complete grid structure regarding typical households which include usual appliances, consumers, battery storage units, and renewable energy sources as generation units is shown in **Figure 14**.

Some of the DC microgrids worth mentioning with ring or interconnected configurations are deployed as follows: Bosch DC Microgrid at California

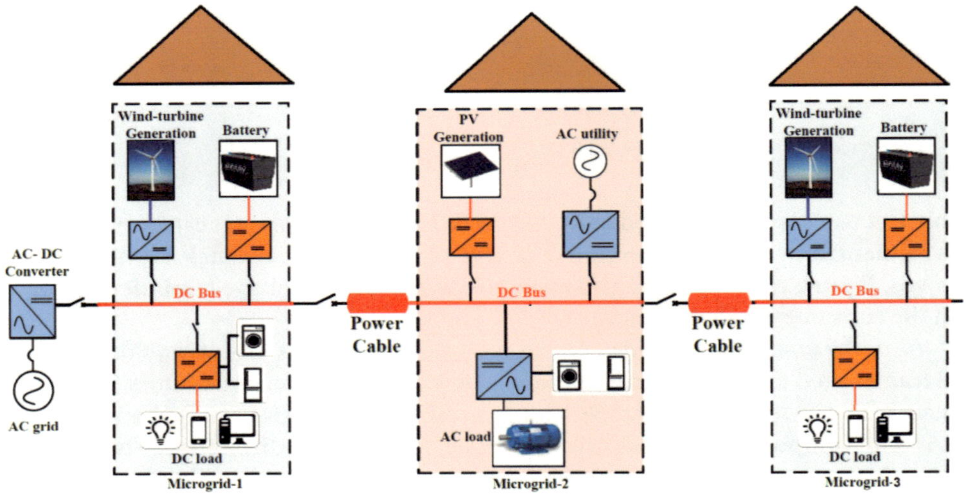

Figure 12.
Radial DC grid topology [22].

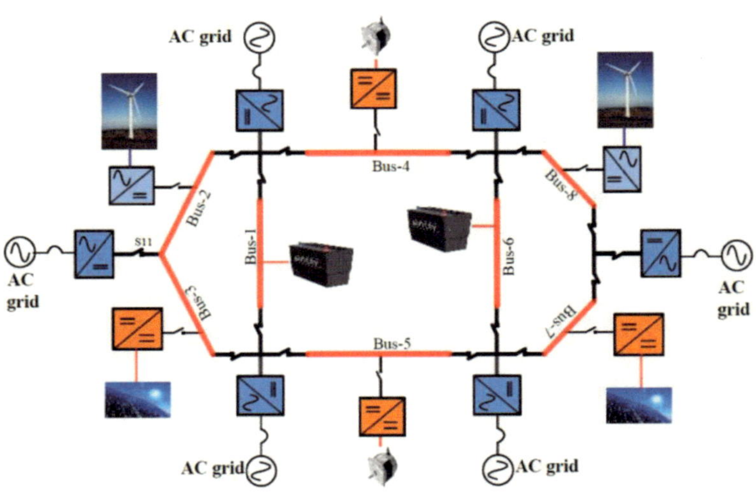

Figure 13.
Annular interconnected DC grid [22].

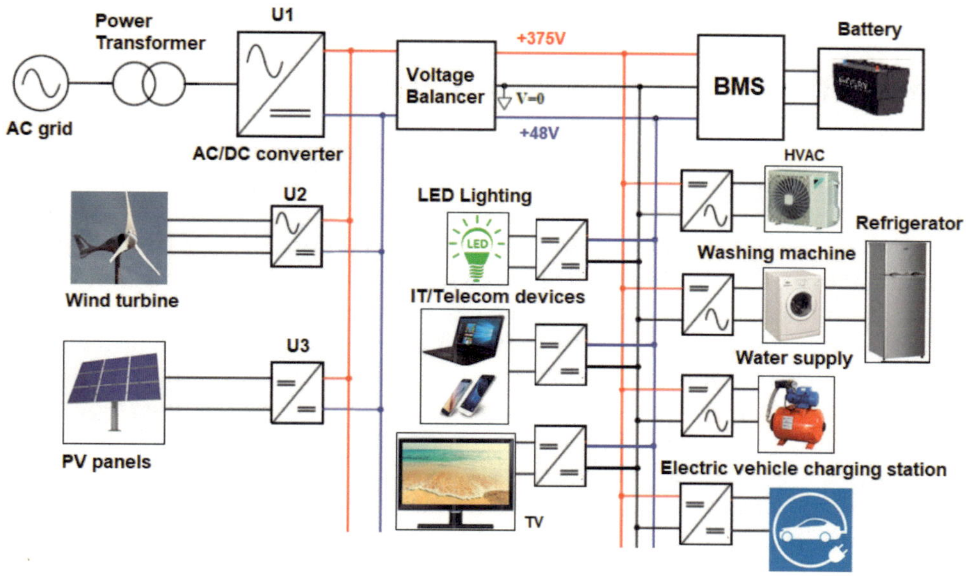

Figure 14.
Topology of a household DC power distribution system which integrates typical consumers as well as renewable energy sources.

Honda Facility; Burlington DC Microgrid Canada, North America, Ontario; Xiamen University DC Microgrid China; Intelligent DC Microgrid Living Lab Denmark.

Depending on the costs and consumer requirements, the DC layout which is outlined above may vary. When taking into account the perspective of renewable energy production, the variation along with the site's weather conditions must be considered. Moreover, the predicted power varies over time, throughout the day, or depending on the season. Thus, the connection to the public distribution grid is necessary if available. Storage systems also allow the accumulation of surplus energy when consumption is low in relation to production capacity.

5. Conclusions

The current chapter envisages and analyzes both advantages and several topologies of LVDC distribution systems for residential applications.

Various analyses regarding energy savings and voltage levels while addressing LVDC distribution systems have been presented. Studies have shown that DC grids will increase energy efficiency and power quality. There are still solutions to find to the challenges that DC grid systems have to overcome compared to existing AC systems, such as equipment safety and protection.

LVDC distribution systems provide particular advantages when integrating renewable energy sources along with storage systems. The development of DC power solutions designed for on the market household appliances may be the next step for the promotion of LVDC systems, especially when minor changes are required. Isolated areas without access to public electricity grid can opt for energy generation from renewable sources. In this case, LVDC grids are the first energy distribution options given the low deployment costs.

For the time being, the home appliance industry is mainly focused on AC power supplied products. There is an increasing number of DC-powered equipment that uses switching mode power supplies for AC conversion as well as for voltage level regulation. These devices can be easily modified by eliminating the rectifying and power factor correction stages.

Most switching mode power sources are already very efficient even with DC/AC conversion module included. Therefore, imposing a different power grid supply for households must be justified especially by the advantages related to the energy production and distribution stages and less by the aspects which regard the final consumer.

It can be concluded that modifying switching home appliances to DC power configurations can be achieved without any significant problems or at significant costs. The transition is easier when using new equipment that mainly embeds switching power supplies or *Inverter* technology to drive washing machine motors or compressors for refrigerators and air conditioning systems.

Older generation equipment that still uses transformers or AC motors require more significant changes of the power sources or the use of DC/AC converters.

The lack of both commercially available electronic devices and systems, and their required standards and regulations, is a major challenge that hinders the rapid development of this field. Currently, it is difficult to identify converters, fuses, or chargers which are necessary for the implementation of DC systems as well as to design and implement a DC microgrid, especially when operating at different voltage levels.

Moreover, potential users do not yet have access to home appliances provided with DC power options. In terms of developing standards, several international organizations such as Emerge Alliance, ETSI-IEC, and IEEE are taking steps in order to establish regulations which are needed for the implementation of DC systems designed for residential applications.

Even though the DC grid system protection is more difficult to be achieved than in the case of AC systems, a corresponding selection of a grounding configuration can be established by using adequate protection devices.

Acknowledgements

This work was supported by a grant of the Romanian Ministry of Research and Innovation, CCCDI-UEFISCDI, project number PN-III-P1-1.2-PCCDI-2017-0391/CIA_CLIM—Smart buildings adaptable to the climate change effects, within PNCDI III.

Also, the authors acknowledge the financial support of MCI through the contracts no. 35N/2018 and 30PFE/2018 between the National R&D Institute for Electrical Engineering ICPE-CA and Romanian Ministry of Research and Innovation (MCI).

Conflict of interest

The author declares that there is no conflict of interest. Thus, there are no conflicts of interest to disclose.

Author details

Lucia-Andreea El-Leathey
National Institute for Research and Development in Electrical Engineering
ICPE-CA, Bucharest, Romania

*Address all correspondence to: andreea.elleathey@icpe-ca.ro

References

[1] Chauhan RK, Rajpurohit BS, Pindoriya NM. DC power distribution system for rural applications. In: 8th National Conference on Indian Energy Sector Synergy with Energy; 11-12 October 2012. AMA Ahmedabad; 2012. pp. 108-112

[2] Koldby E, Hyttinen M. Challenges on the road to an offshore HVDC grid. In: Nordic Wind Power Conference; 10-11 September 2009; Bornholm, Denmark. 2009

[3] Rudervall R, Charpentier JP, Raghuveer S. High Voltage Direct Current (HVDC) Transmission Systems Technology Review Paper. Energy Week; 7-8 March 2000; Washington, D.C, USA; 2000

[4] Why HVDC. Economic and Environmental Advantages [Internet]. Available from: https://new.abb.com/systems/hvdc/why-hvdc/economic-and-environmental-advantages [Accessed: December 21, 2018]

[5] Eltamaly AM, Elghaffar ANA. HVDC system control between different frequencies networks and fault analysis with HVAC system. In: International Egyptian Engineering and Technology Journal; IEETJ No: 000-2017. 2017. pp. 1-9

[6] Moreno AF, Mojica-Nava E. LVDC microgrid perspective for a high efficiency distribution system. In: IEEE PES Transmission & Distribution Conference and Exposition-Latin America (PES T&D-LA); 10-13 September 2014; Medellin, Colombia. 2014. DOI: 10.1109/TDC-LA.2014.6955283

[7] Justo J, Mwasilu F, Lee J, Jung J. AC-microgrids versus DC-microgrids with distributed energy sources: A review. Renewable and Sustainable Energy Reviews. 2013;**24**:387-405

[8] Elsayed AT, Mohamed AA, Mosammed OA. DC microgrids and distribution systems: An overview. Electric Power Systems Research. 2015;**119**:407-417

[9] Mohamed A, Ghareeb A, Youssef T, Mohammed OA. Wide area monitoring and control for voltage assessment in smart grids with distributed generation. In: IEEE PES Innovative Smart Grid Technologies Conference (ISGT); 24-27 February 2013; Washington, D.C, USA. 2013. DOI: 10.1109/ISGT.2013.6497849

[10] El-Leathey LA. Energy management system designed for the interconnected or islanded operation of a microgrid using LabVIEW software. In: Smart Microgrids. Rijeka, Croatia: IntechOpen; 2018. pp. 45-64. DOI: 10.5772/intechopen.74856

[11] Romanian Standards Association ASRO [Internet]. 2018. Available from: www.asro.ro

[12] Energy Regulations. IEC 60898-1 and IEC 60947-2: A Tale of Two Standards [Internet]. Available from: https://blog.schneider-electric.com/power-management-metering-monitoring-power-quality/2013/07/16/iec-60947-2-the-all-risk-insurance-for-circuit-breakers/ [Accessed: December 21, 2018]

[13] Resi9. Consumer Unit and Plug in Circuit Protection, with the Highest Levels of Residential Circuit Protection Safety [Internet]. Available from: https://www.schneider-electric.co.uk/en/product-range/61364-resi9/ [Accessed: December 21, 2018]

[14] Pritchard E, Gregory DC, Srdic S. The DC revolution. IEEE Electrification Magazine. 2016;**4**:9

[15] Power Factor Correction in Switching Mode Power Supplies [Internet].

Available from: http://electronica-azi.ro/2012/07/03/corectia-factorului-de-putere-in-sursele-de-alimentare-smps/ [Accessed: December 21, 2018]

[16] Features of Mitsubishi Electric Power Modules for Home Appliances [Internet]. Available from: http://www.mitsubishielectric.com/semiconductors/application/home/index.html?seriesid=09 [Accessed: December 21, 2018]

[17] DC/DC Converters [Internet]. Available from: http://electronica-azi.ro/2011/02/01/convertoare-dcdc/ [Accessed: December 21, 2018]

[18] Wunder B, Ott L, Szpek M, Boeke U, Weiß R. Energy efficient DC-grids for commercial buildings. In: IEEE 36th International Telecommunications Energy Conference (INTELEC); Vancouver, BC. 2014. pp. 1-8. DOI: 10.1109/INTLEC.2014.6972215

[19] Weiss R, Ott L, Boeke U. Energy efficient low-voltage DC-grids for commercial buildings. In: IEEE First International Conference on DC Microgrids (ICDCM); Atlanta, GA. 2015. pp. 154-158. DOI: 10.1109/ICDCM.2015.7152030

[20] Eremia M, Toma L. Către orașele inteligente ale viitorului—Smart Cities. In: Towards Future Smart Cities. 7th Annual ASTR Conference. 2012

[21] 2016-2030 Romania's Energy Strategy with 2050 Perspectives [Internet]. Available from: http://www.mmediu.gov.ro/app/webroot/uploads/files/2017-03-02_Strategia-Energetica-a-Romaniei-2016-2030.pdf [Accessed: December 21, 2018]

[22] Rodriguez-Diaz E, Chen F, Vasquez JC, Guerrero JM, Burgos R, Boroyevich D. Voltage-level selection of future two-level LVdc distribution grids. IEEE Electrification Magazine. 2016;**4**(2):20-28. DOI: 10.1109/MELE.2016.2543979

Section 2

Operation, Control and Optimization

Chapter 4

Power Quality Improvement of a Microgrid with a Demand-Side-Based Energy Management System

Gaspard d'Hoop, Olivier Deblecker and Dimitrios Thomas

Abstract

This chapter addresses the power quality of grid-connected microgrids in steady state. Three different power quality issues are evaluated: the voltage drop, the harmonic distortion, and the phase unbalance. A formulation for an energy management algorithm for microgrids is proposed under the form of a mixed-integer linear optimization including harmonic load flows. It handles both the optimization of the scheduling of all the generation, storage, and load assets, and the resolution of power quality issues at the tertiary level of control by adjusting the levels of certain types of loads within the system. This algorithm is simulated for different scenarios on a conceptual test-case microgrid with residential, industrial, and commercial loads. The results show that the demand-side management mechanism inside the algorithm can adapt efficiently the consumption behavior of certain loads, so that the voltage drop, the voltage total harmonic distortion, and the voltage unbalance factor meet the required standards at every node of the microgrid during the day. It is also highlighted that the microgrid can gradually reduce the purchase of power from the utility grid to which it is connected if the electricity price on the spot market increases.

Keywords: microgrid, power quality, demand-side management, energy management system, mixed-integer linear programming

1. Introduction

The sectors of electricity generation, transmission, and distribution are currently facing a change of paradigm. The introduction of decentralized renewable energy sources and storage systems, the rise of electric vehicles, the multiplication of international high-voltage lines, and the development of smart grids are many reasons to believe that the way our societies produce, transmit, and consume electricity will progressively change in the coming decades [1].

A reliable and economically feasible supply of electricity remains, however, primordial for industrial and residential consumers. In the near future, several countries plan to introduce real-time pricing (RTP) at distribution levels, in order to reflect directly the variability of the electricity price on the consumers [2]. But since the costs of distributed renewable sources and storage systems have been decreasing, many important consumers see an opportunity to build their own microgrid. Their goal is to reduce their electricity bills by covering partially or totally their electrical loads [3]. Buildings, factories, or even residential neighborhoods are often referred to as the types of consumers that could foresee the creation of a microgrid if it appears to be economically viable [4]. In areas where blackouts and grid outages are frequent, certain microgrids have the interesting capability to isolate themselves from the rest of the grid, and thus act as an uninterrupted power supply (UPS) for the sensitive loads. This islanding mode is also called *stand-alone mode* of a microgrid, in opposition to the *grid-connected mode*.

The presence of nonlinear, single-phase, or highly inductive loads in the system can impair the usual three-phase direct symmetric voltage and current waveforms. In general, the parameters concerning reliability and waveform quality of the voltage and current are part of the so-called *power quality* (PQ), a set of characteristics which define an adequate supply of electricity [5]. According to the performance required and the operative standards, specific bounds should be defined for each PQ index. It is moreover essential for microgrids, as several scientific articles have pointed out that PQ-related issues are more frequent in this type of architecture, especially when they are disconnected from the main grid [6].

For larger networks, PQ issues are normally handled with different techniques at the primary level of control, such as droop control of synchronous machines, active filtering, static VAR compensators (SVC), or other kinds of equipment [7]. But the cost of most of these installations is high for small-scale microgrids. Hence, another less effective but more economical method to tackle long-lasting PQ issues consists in regulating the demand levels of flexible loads to stabilize the microgrid [6]. This modification of the energy used by the consumers is called *demand-side management* (DSM) and requires an adapted communication system between the consumers and the central controller of the microgrid.

This DSM feature is included inside the energy management system (EMS) algorithm, situated at the tertiary level of control of the microgrid [8]. The EMS is normally only focused on dispatching the active and reactive power fluxes between the distributed energy resources (DERs), the utility grid, and the loads to satisfy the active and reactive power balances while reaching an economic optimum on a daily basis [5]. However, including an appropriate DSM mechanism inside the EMS makes it also a potential solution for the assessment of the PQ in steady state. To our knowledge, very few authors have proposed such an approach based on PQ-related scheduling decisions taken over the multihour horizon. In [6], an energy scheduling algorithm is presented, aiming at mitigating PQ issues through coordinating the operating schedules of sensitive devices in a commercial building microgrid. Yet, most works in literature are based on intelligent control strategies of the DERs interfacing inverters or utilization of dedicated power electronic devices for compensation, all acting on a faster time scale (see, e.g., [9] for a thorough survey of PQ improvement techniques in microgrids).

The main objective of the present chapter is to design and simulate an EMS algorithm for microgrids with specific PQ-related constraints while analyzing its behavior on realistic situations. The resulting algorithm proposed in this work investigates three different PQ issues, namely voltage drop, phase unbalance, and harmonic distortion. It also introduces new considerations regarding the operation of diesel generators during the transitions between grid-connected and stand-alone modes. The main advantage of the proposed algorithm is that it can effectively manage the abovementioned issues, whether the primary control means for PQ enhancements are available or not in the microgrid. The chapter starts with a description of the PQ indices and the common recommendations for their evaluation. Then, the model chosen for the DSM of the loads is presented in the following section. The third section enlightens the equations governing the EMS algorithm and outlines their role in the whole control mechanism of the microgrid. The fourth section presents and analyzes the results of the simulations for different scenarios on a test-case microgrid. A conclusion ends this chapter with the reminders on the conceptual approach, the procedure, and the major findings, while pointing the way toward further research.

2. Power quality issues in steady state

The term power quality has often been the subject of different interpretations. In [10], e.g., the definition encompasses several aspects without giving defined bounds: "Electric power quality is a term that refers to maintaining the near sinusoidal waveform of power distribution bus voltages and currents at rated magnitude and frequency. Thus, PQ is often used to express voltage quality, current quality, reliability of service, quality of power supply, etc."

In this chapter, the attention will be focusing on issues with a large time scale (at every hour), such as undervoltage, voltage phase unbalance, and voltage harmonic distortion.

2.1 Voltage deviation

The voltage magnitude at a bus can deviate from its rated value. These deviations are often tolerated for small percentages, but if they cross certain limits, they are considered as disturbances. Considering a short line model between nodes 1 and 2, the real part of the voltage difference between the two nodes ΔV_d is given by the well-known expression [11]:

$$\Delta V_d = \frac{RP_1 + XQ_1}{V_1} \cong V_1 - V_2 \qquad (1)$$

where P_1 and Q_1 are the active and reactive power flowing from node 1 to node 2.

Eq. (1) represents the voltage drop across the line. It clearly shows that it is highly dependent on the reactive power flow for inductive lines. However, in distribution lines with lower reactance on resistance ratio (X/R) such as in microgrids, both the active and reactive powers have an impact on the voltage deviation. The IEEE 1547-2003 standard recommends that a microgrid should not make voltage variations greater than $\pm5\%$ around the nominal value [12]. The new norm IEEE 1547.4 also recommends that at least one DER should be responsible for regulating the voltage and the frequency in islanded mode, while staying in coordination with the other loads and DERs. In this chapter, the attention is focused on voltage drop.

2.2 Harmonic distortion

In recent years, the rising use of nonlinear power electronic devices along with an increase of the sensitive loads has resulted in various concerns [13]. The continued presence of harmonics can damage or degrade components in the networks such as transformers, electric motors, or electronic appliances. Harmonics also increase the total amount of power losses in the system [13]. High harmonics can normally be easily filtered by passive or active filters or, at least, reduced using appropriate modulation schemes in the control of the power electronic switches. However, low harmonics (3rd, 5th, 7th, 11th, etc.) are difficult to filter without reducing in the same way the signal at the base frequency. Harmonic cancellation techniques exist to tackle this problem, but they are usually expensive and technically difficult to implement [14].

The most used index to evaluate the distortion of a signal is called the total harmonic distortion (THD). The correct definition according to the IEC 61000-2-2 standard states that it is the ratio, in percentage, of the root-sum-square of all the harmonic magnitudes (without including the fundamental) on the magnitude of the fundamental [15]. THDs for voltage and current can thus be written as:

$$THD_V = \frac{\sqrt{\sum_{h=2}^{N} V_h^2}}{V_1}, THD_I = \frac{\sqrt{\sum_{h=2}^{N} I_h^2}}{I_1} \quad (2)$$

At low voltages, the limit of 5% voltage THD is commonly used [16]. In this work, the current THD will be considered as a known factor for each load. But the voltage THD, which results from those current harmonics, will be the PQ index measured in each node and the one that will be effectively constrained with the bound of 5%.

2.3 Phase unbalance

According to the theory of the Fortescue decomposition, any three-phase system can be decoupled into direct, inverse, and zero-sequence components. The voltage unbalance is generally evaluated with an index called voltage unbalance factor (VUF), which corresponds to the ratio between the inverse (V_i) (or zero (V_h)) and direct-sequence (V_d) components of the Fortescue decomposition of the voltage:

$$VUF = max\left(\frac{V_i}{V_d}, \frac{V_h}{V_d}\right) \quad (3)$$

The IEEE 1547.4-2011 standard warns the owner of the microgrid that large voltage unbalances can cause problems to the three-phase inverter-based DERs, by placing high ripple currents on the DC bus. These ripple currents may also have an adverse effect on the synchronous generators and energy sources (like batteries and fuel cells). The norm recommends the objective of keeping a VUF lower than 3% at every node [12].

3. Building the energy management system

3.1 Control structure of an AC microgrid

One can easily understand that communication between all the actors of the microgrid is essential, in order to maintain satisfying performance and to minimize the operating costs in both grid-connected and stand-alone modes. The control of AC microgrids is generally divided into a hierarchical three-level structure. In a similar way to traditional networks, the levels are differentiated by their usual time response. As shown in **Figure 1**, the entity that will supervise the communication between the different layers is called *microgrid central controller* (MGCC) [17].

The EMS algorithm supporting PQ that this work aims to develop is part of the tertiary control scheme. The algorithm will essentially compute cost minimizing

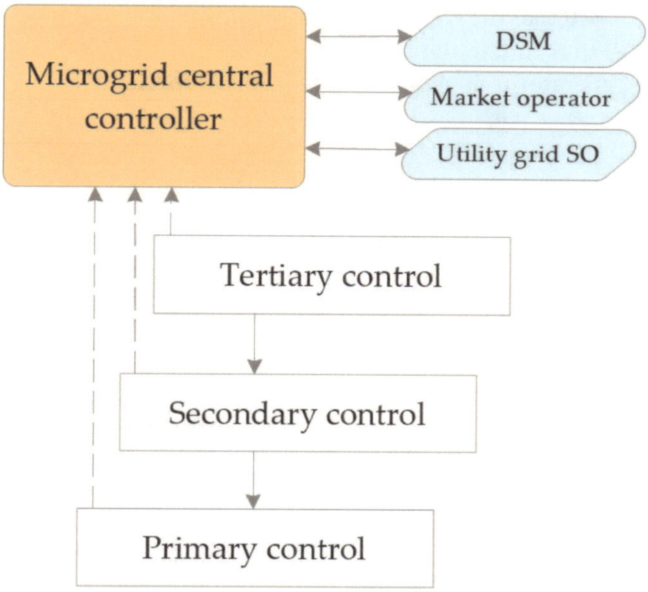

Figure 1.
Hierarchical control structure in microgrids (dashed lines represent feedback signals and solid lines represent guideline information).

optimization for 24 h, with specific decisions at each hour, thanks to a mixed-integer linear programming (MILP) algorithm. The algorithm will set the operating point of the production, storage systems, and load levels at every hour and assumes inherently that the frequency remains constant at 60 Hz. Steady-state PQ issues are addressed as well under the form of constraints, namely voltage drop, phase unbalance, and harmonic distortion. As already mentioned, it is interesting to notice that in traditional distribution grids, those PQ issues would normally be treated at the primary or secondary level of control, often with static compensator or dynamic power-electronics-based devices. But those installations are often too expensive for small-scale microgrids.

The constraints of the EMS algorithm can be divided into different sections, depending on the asset they concern: loads, grid connection, DGs, ESSs, etc. They are presented below using the following indices notations:

- The time index in hours is called t (1, ..., 24).
- The phase index is called ph (a, b, c).
- The harmonic order index is called h (1, 3, 5, ..., 11).
- The load index is called l ($Load_1$, $Load_2$, etc.).
- The load type index is called ty (HVAC, appliances, motors, etc.).
- The DG index is called r ($genset_1, genset_2$, PV, WT, etc.).
- The genset index is called r_{genset} ($genset_1, genset_2$, etc.).
- The ESS index is called es (es_1, es_2, etc.).
- The iteration index of the PQ regulation loop is denoted it (0, 1, 2, etc.).

It should be noted that the r_{genset} index gathers the different diesel generators that will have the task to maintain the frequency during stand-alone mode. Practically, only one diesel generator is enough if its capacity is sufficient.

3.2 Loads

3.2.1 Load-related parameters and variables

To represent the diversity of loads present in the system, each aggregated load l at a node of a microgrid is disaggregated into several subloads depending on the end use, called *types of load* and denoted *ty*. These types of loads gather common devices that have similar power factor, harmonic content, and flexibility, namely HVAC, domestic hot water (DHW), lights, appliances, and motor drives. The purpose of this distinction is to get a better knowledge of the types of equipment that are causing PQ issues within the aggregated loads. It could be, for example, foreseeable that the MGCC asks a consumer at a certain point of the network to reduce slightly the temperature of its building to avoid the purchase of power on the utility grid when this one is expensive.

The parameter $type_{ratio}(l, ty)$ tells us the usual proportion of a load type *ty* in the consumption of an aggregated load l. For example, the $type_{ratio}(l, ty)$ of motors inside an industrial load is usually higher than 0.5. Each type of load also has the following parameters: fundamental-frequency power factor ($pf(ty)$), harmonic content ($harm(ty, h)$), current THD ($THD_I(ty)$), and flexibility ($flex(ty)$), i.e., the percentage of the power consumption that a device can reduce without overly affecting its users. Those parameters are considered to be constant for every load type, whatever the active power they consume.

At an aggregate level, the total demand curve of a load at a specific node of the microgrid is supposed to be known and is referenced as $P_{load,\,curve}(l, t)$. Note that it is evaluated at the normal consumption usage, without the possible reduction linked with DSM. The aggregated loads could also be unbalanced between their phases. The phase distribution index $phase_{distrib}(l, ph)$ represents the ratio between the power that is actually required by a phase (a, b, or c) of an aggregated load and the power that would normally be consumed by this phase if the load was balanced. Thus, if these indices are unitary for all the phases of a load, this load is perfectly balanced. Finally, another parameter that will be useful for PQ matters is $x/r(l)$, the ratio between reactance and resistance of the line to which the load is connected.

At each hour, each load type can be either active or not, this binary variable is called $On_{load}(l, ty, t)$ and is equal to 1 if the load is active. If it is the case, the active and reactive powers that the type of load effectively consumes are called $P_{load}(l, ty, t)$ and $Q_{load}(l, ty, t)$. The total active and reactive powers for the aggregated loads at each node of the microgrid are called $P_{load,\,tot}(l, t)$ and $Q_{load,\,tot}(l, t)$.

3.2.2 Load-related constraints

The active power must be between the allowable minimum and maximum if a load type of an aggregated load is activated (i.e., $On_{load}(l, ty, t)$ is equal to 1):

$$\forall l,\ \forall ty,\ \forall t\ P_{load}(l, ty, t) \geq (1 - flex(ty)) \cdot type_{ratio}(l, ty) \cdot P_{load,\,curve}(l, t) \cdot On_{load}(l, ty, t) \quad (4)$$

$$P_{load}(l, ty, t) \leq type_{ratio}(l, ty) \cdot P_{load,\,curve}(l, t) \cdot On_{load}(l, ty, t) \quad (5)$$

DOI: http://dx.doi.org/10.5772/intechopen.83604

The reactive power consumption is calculated using the active power and the fundamental-frequency power factor ($pf = cos(\phi)$):

$$\forall l, \forall ty, \forall t \ Q_{load}(l, ty, t) = P_{load}(l, ty, t) \frac{\sqrt{1 - pf^2(ty)}}{pf(ty)} \tag{6}$$

Finally, the aggregated load can be calculated for every node as follows:

$$\forall l, \forall t \ P_{load, tot}(l, t) = \sum_{ty} P_{load}(l, ty, t) \tag{7}$$

$$Q_{load, tot}(l, t) = \sum_{ty} Q_{load}(l, ty, t) \tag{8}$$

3.3 Distributed generators

The DGs are generally interfaced with the grid by a fully controllable converter. These converters have the ability to absorb or generate reactive power in addition to the active power generation. Since the converter is independent from the DG, it can absorb or generate reactive power, even if the DG is turned off. The left-hand graph in **Figure 2** shows the capability curve of the DG converter with active and reactive power axes. Normally, the rating of power electronic converters is expressed in VA, which means that the feasible zone should be a circle on an active and reactive power axis diagram. However, linear programming does not allow quadratic equations and the capability zone of the converter is then approximated by a rectangle. The absolute reactive power remains anyway small in general and this simplification would normally not affect significantly the results.

3.3.1 DG-related parameters and variables

If the DG is on, it will be able to regulate its output useful power between a maximum and a minimum. Of course, renewable and intermittent sources are less flexible concerning this last feature because they often depend on meteorological conditions. The minimal and maximal active power outputs if the DG is running are denoted $P_{DG, \min}(r, t)$ and $P_{DG, max}(r, t)$. The absolute reactive power generated or absorbed by the converter must also be enclosed by a maximum $Q_{DG, max}(r, t)$, depending on its rating.

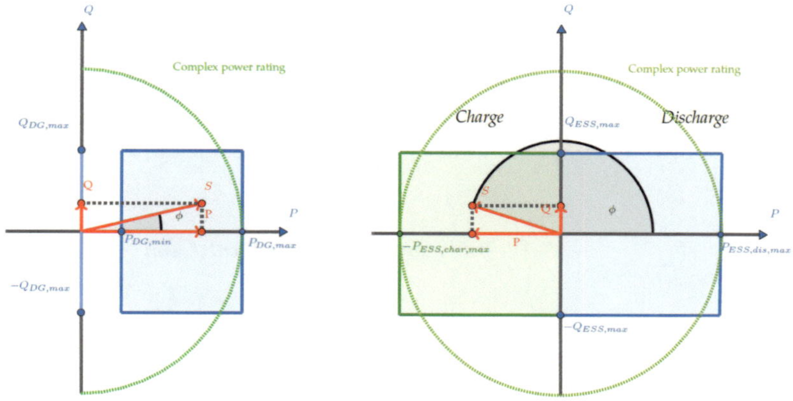

Figure 2.
Capability zone of a DG and an ESS converter, with active and reactive power feasible ranges (in blue for discharging and green for charging). The real power rating (in VA) of the power electronic converter is represented with a dashed green line.

As for the loads, the DGs in the system can either be activated or not, depending on if they produce effectively active power or not. This binary variable is called $On_{DG}(r,t)$. The output active and reactive powers are denoted $P_{DG}(r,t)$, $Q_{DG,gen}(r,t)$, and $Q_{DG,abs}(r,\ t)$. The start-up and shutdown of a DG at a specific hour are also binary variables and are denoted $ST_{DG}(r,t)$ and $SD_{DG}(r,t)$. Finally, the $offgrid(t)$ binary variable is equal to 1 if the microgrid operates in stand-alone mode at the hour t.

3.3.2 DG-related constraints

If the DG is activated (the variable $On_{DG}(r,t)$ is equal to 1), the active power must be between the allowable minimum and maximum:

$$\forall r, \forall t\ P_{DG,min}(r,t)\cdot On_{DG}(r,t) \le P_{DG}(r,t) \le P_{DG,max}(r,t)\cdot On_{DG}(r,t) \tag{9}$$

On the opposite, a DG can produce or absorb reactive power even if the DG itself does not produce active power, it then behaves like a dynamic VAR compensator:

$$\forall r, \forall t\ Q_{DG,gen}(r,t) \le Q_{DG,max}(r,t) \tag{10}$$

$$Q_{DG,abs}(r,t) \le Q_{DG,max}(r,t) \tag{11}$$

In stand-alone mode, the role of certain diesel generators (*gensets*) is to regulate the frequency. For this reason, they should not directly contribute to the reactive power balance, but the converters of other DERs can take this task.

$$\forall r_{genset}, \forall t\ Q_{DG,gen}(r_{genset},t) \le Q_{DG,max}(r_{genset},t)\cdot(1-offgrid(t)) \tag{12}$$

$$Q_{DG,abs}(r_{genset},t) \le Q_{DG,max}(r_{genset},t)\cdot(1-offgrid(t)) \tag{13}$$

The start-up and shutdown of a DG at the hour t are defined as follows:

$$\forall r, t\in\{2...24\}\ ST_{DG}(r,t) - SD_{DG}(r,t) = On_{DG}(r,t) - On_{DG}(r,t-1) \tag{14}$$

$$ST_{DG}(r,t) + SD_{DG}(r,t) \le 1 \tag{15}$$

Another novel constraint added to this algorithm is the use of diesel generator engines (*genset*) to smooth the transition between grid-connected and stand-alone modes. Indeed, in grid-connected mode, the stabilization of the frequency is achieved by the main utility grid. But in stand-alone mode, the diesel generators included in r_{genset} (at least one) will have to take over the task. If the binary variable *offgrid* (t) is equal to 1 at the hour t, the constraint then forces at least one *genset* to produce active power during each hour of stand-alone operation, and 1 h before and 1 h after as well.

$$\forall t, \forall r_{genset}\ offgrid(t) \le \frac{1}{3}\begin{pmatrix} On_{DG}(r_{genset},t-1) + On_{DG}(r_{genset},t) \\ +On_{DG}(r_{genset},t+1) \end{pmatrix} \tag{16}$$

3.4 Energy storage systems

As it is shown on the right-hand graph in **Figure 2**, the power converters interfacing the energy storage systems with the microgrid are usually able to work in the four quadrants of the active and reactive power plane. This is explained by the fact that they can manage a bidirectional flow of active and reactive powers.

Note that the generator convention is adopted, so, e.g., the active power is positive when discharging. As it was already explained in the previous section, the capability zone of these converters is approximated by a rectangle instead of a circle, due to the limitations of linear programming.

3.4.1 ESS-related parameters and variables

Every storage system in the microgrid possesses the following parameters: round-trip efficiency $eff(es)$, maximal useful energy $E_{ESS}(es)$, maximal reactive power $Q_{ESS,max}$, and maximal power for charging $P_{ESS,char,max}(es)$ and discharging $P_{ESS,dis,max}(es)$. Another information needed is the initial state of charge (SoC) of the storage system at the hour preceding the beginning of the simulation, denoted $SOC_{init}(es)$ [18].

The different variables for each ESS are the amount of active power it delivers or it absorbs, $P_{ESS,dis}(es,t)$ and $P_{ESS,char}(es,t)$, the amount of reactive power it generates or absorbs, $Q_{ESS,gen}(es,t)$ and $Q_{ESS,abs}(es,t)$, and the remaining SoC at every hour $SOC(es,t)$. The binary variables $ESS_{char}(es,t)$ and $ESS_{dis}(es,t)$ are equal to 1 if the ESS charges or discharges active power, respectively.

3.4.2 ESS-related constraints

The first constraint forces the SoC of the storage system to be between 0 and 100%:

$$\forall es, \forall t \; 0 \le SOC(es,t) \le 1 \tag{17}$$

At hour 1, the ESS needs a special equation because it uses the initial SoC parameter:

$$\forall es, t = 1 \; SOC(es,t) = SOC_{init}(es) + \frac{P_{ESS,char}(es,t) \cdot \sqrt{eff(es)}}{E_{ESS}(es)} - \frac{P_{ESS,dis}(es,t)}{\sqrt{eff(es)} \cdot E_{ESS}(es)} \tag{18}$$

The equation for the remaining hours has the same form as the previous one, except that $SOC_{init}(es)$ is replaced by $SOC(es, t-1)$. So, $\forall es, t = 2...24$:

$$SOC(es,t) = SOC(es,t-1) + \frac{P_{ESS,char}(es,t) \cdot \sqrt{eff(es)}}{E_{ESS}(es)} - \frac{P_{ESS,dis}(es,t)}{\sqrt{eff(es)} \cdot E_{ESS}(es)} \tag{19}$$

The following inequalities restrict the active power of the charge and discharge by their maximal bounds:

$$\forall es, \forall t \; P_{ESS,char}(es,t) \le P_{ESS,char,max}(es) \cdot ESS_{char}(es,t) \tag{20}$$

$$P_{ESS,dis}(es,t) \le P_{ESS,dis,max}(es) \cdot ESS_{dis}(es,t) \tag{21}$$

The same constraints as for the DGs apply concerning the generation and absorption of reactive power:

$$\forall es, \forall t \; Q_{ESS,gen}(es,t) \le Q_{ESS,max}(es) \tag{22}$$

$$Q_{ESS,abs}(es,t) \le Q_{ESS,max}(es) \tag{23}$$

Finally, it seems natural that a storage system cannot charge and discharge its active power at the same time:

$$\forall es, \forall t \; ESS_{char}(es,t) + ESS_{dis}(es,t) \leq 1 \tag{24}$$

3.5 Power exchange with the utility grid

As long as it is not in stand-alone mode, the microgrid is physically connected to a larger traditional utility grid at the point of common coupling (PCC).

3.5.1 Grid-related parameters and variables

The maximal power that can be drawn from the grid is denoted $P_{grid,\,in,\,max}$. If the microgrid produces a surplus of energy, it can inject it into the grid. The maximum power that can be injected is called $P_{grid,\,out,\,max}$. In grid-connected mode, the microgrid can also help the utility grid to achieve its own voltage stability by providing the reactive power it needs, which is denoted $Q_{grid,\,req}(t)$.

The variables concerning the connection with the utility grid are the active power drawn from it, $P_{grid,\,in}(t)$, and rejected to it, $P_{grid,\,out}(t)$, and the reactive power absorbed from it, $Q_{grid,\,in}(t)$, or supplied to it, $Q_{grid,\,out}(t)$.

3.5.2 Grid-related constraints

The active and reactive power can, of course, only be exchanged if the switch at the PCC is closed, and so if the binary variable *offgrid*(t) is equal to zero:

$$\forall t \; P_{grid,\,in}(t) \leq P_{grid,\,in,\,max} \cdot (1 - offgrid(t)) \tag{25}$$

$$P_{grid,\,out}(t) \leq P_{grid,\,out,\,max} \cdot (1 - offgrid(t)) \tag{26}$$

$$Q_{grid,\,in}(t) \leq Q_{grid,\,in,\,max} \cdot (1 - offgrid(t)) \tag{27}$$

$$Q_{grid,\,out}(t) \leq Q_{grid,\,out,\,max} \cdot (1 - offgrid(t)) \tag{28}$$

3.6 Power balance

Some of the most essential equations in the algorithm are the balance equations that will equilibrate load and generation. Here are the equations for the active and reactive power balances:

$$\begin{aligned} \forall t \; \textstyle\sum_l (P_{load}(l,t) \cdot (1 + loss_P)) + \sum_{es} P_{ESS,\,char}(es,t) \\ + P_{grid,\,out}(t) = \textstyle\sum_{es} P_{ESS,\,dis}(es,t) + \sum_r P_{DG}(r,t) + P_{grid,\,in}(t) \end{aligned} \tag{29}$$

$$\begin{aligned} \forall t \; \textstyle\sum_l (Q_{load}(l,t) \cdot (1 + loss_Q)) + \sum_{es} Q_{ESS,\,abs}(es,t) + \sum_r Q_{DG,\,abs}(r,t) \\ + Q_{grid,\,out}(t) + Q_{grid,\,req}(t) \cdot (1 - offgrid(t)) = \textstyle\sum_{es} Q_{ESS,gen}(es,t) \\ + \textstyle\sum_r Q_{DG,\,gen}(r,t) + Q_{grid,\,in}(t) \end{aligned} \tag{30}$$

In this work, predetermined hourly values are used for $loss_P$ and $loss_Q$ that represent the upper bounds of losses in the system, as a ratio to the total active or reactive power demand. These values are calculated by the load flow studies and represent a very small percentage of the total hourly loads.

3.7 Power quality issue avoidance

The PQ constraints are not directly active at the first dispatch optimization of the MILP algorithm. The results will be first analyzed with a load flow and only then, if some standards are violated, new constraints will be added to address these issues. One difficulty is that the MILP optimization of the EMS only gives active and reactive power results and not voltages. It is therefore not possible to compute the voltage drop, the voltage THD, and the VUF directly. Three new constraints are thus defined to reflect the PQ issues, but with the leverage of active and reactive powers. The aim of this approach is to act on the demand level of problematic types of loads at the different nodes of the microgrid. For instance, if the voltage THD is above the limits at a node with residential loads, then the algorithm will ask the customers to reduce the consumption of their nonlinear loads, within a feasible range, to support the microgrid PQ and help to secure the safety of their own electrical equipment.

3.7.1 Power-quality-related parameters and variables

As mentioned previously, multiple iterations are launched between the MILP optimization and harmonic load flows if at least one PQ disturbance exceeds the standards. The goal is to find at the end of the iteration process the right constraints to add in the optimization that will prevent any PQ issue. In this work, sensitivity parameters $\lambda_{\Delta V}$, λ_{dist}, and λ_{unb} are used to strengthen these constraints at every iteration loop. The tuning of these sensitivity parameters essentially results in a trade-off between the accuracy of the final results and the speed of convergence. In practice, the voltage indices (ΔV, THD_V, VUF) should be just below the standard in the best case, to limit the impacts on the levels of the loads.

The variables for the voltage drop, the harmonic distortion, and the phase unbalance in the MILP optimization are denoted $\alpha_{\Delta V}$, α_{dist}, $\alpha_{unb,ab}$, $\alpha_{unb,bc}$, and $\alpha_{unb,ca}$. They will be defined in the equality constraints presented below.

3.7.2 Power-quality-related constraints

The index *it* represents the iteration number. It is initialized at zero and then incremented each time it goes through a harmonic load flow (the variable *i* is in this case a positive integer).

As reminded within the Section 2.1, the voltage drop between two nodes is linked with the active and reactive power flows (see (1)). If the *X/R* ratio of the line that connects each aggregated load is known, then the variable $\alpha_{\Delta V}$ is defined as:

$$\forall l, \forall t \ \alpha_{\Delta V}(l,t)^{it=i} = P_{load,tot}(l,t)^{it=i} + x/r(l) \cdot Q_{load}(l,t)^{it=i} \tag{31}$$

The following equation is applied at every iteration if the voltage is below 95% of its rated value (i.e., ΔV >0.05 pu), at every load bus l and every hour t:

$$\forall l, \forall t \ \alpha_{\Delta V}(l,t)^{it=i+1} \leq \alpha_{\Delta V,\lim}(l,t)^{it=i+1} \tag{32}$$

with:

$$\begin{gathered} \textit{If } |\Delta V(l)| > 0.05\,pu \text{ and } it > 0\text{: } \alpha_{\Delta V,\lim}(l,t)^{it=i+1} \\ = \alpha_{\Delta V}(l,t)^{it=i} \cdot \left(1 - \lambda_{\Delta V} \cdot \Delta V(t)^{it=i}\right) \text{Otherwise: } \alpha_{\Delta V,\lim}(l,t)^{it} = \infty \end{gathered} \tag{33}$$

For the harmonic distortion, Eq. (2) shows that nonlinear loads with a rich harmonic content will have a higher current THD and should therefore be limited when the distortion is too high with respect to the norms. One should notice that in this work, the current THD is a parameter fixed for every type of load ty, while the voltage THD is a PQ index computed with harmonic load flows. The variable α_{dist} represents the total amount of distortion in the network and is defined as:

$$\forall t \ \alpha_{dist}(t)^{it=i} = \sum_l \sum_{ty} \left(P_{load}(l,ty,t)^{it=i} \cdot THD_I(ty) \right) \tag{34}$$

The following equation is applied at every iteration, if the voltage THD is over 5% at least at one load bus l, and at every hour t:

$$\forall t \ \alpha_{dist}(t)^{it=i+1} \leq \alpha_{dist,\lim}(t)^{it=i+1} \tag{35}$$

with:

$$\begin{aligned} &\textit{If } THD_V > 5\% \text{ and } it > 0\text{: } \alpha_{dist,\lim}(t)^{it=i+1} \\ &= \alpha_{dist}(t)^{it=i} \cdot \left(1 - \lambda_{dist} \cdot max_l \left(THD_V(l,t)^{it=i} \right) \right) \text{Otherwise: } \alpha_{dist,\lim}(t)^{it} = \infty \end{aligned} \tag{36}$$

If several buses have an irregular voltage THD at a certain hour, the load bus that has the highest distortion is selected to create $\alpha_{dist,\lim(t)}$ in (36). The constraints on the voltage distortion are global, for the whole microgrid, and only depend on the hour of the day t.

Regarding the phase unbalance, the following set of variables is adopted between each phase:

$$\forall t \ \alpha_{unb,ab}(t)^{it=i} = \left| \sum_l P_{load,tot}(l,t) \cdot (phase_{distrib}(l,a) - phase_{distrib}(l,b)) \right| \tag{37}$$

$$\alpha_{unb,bc}(t)^{it=i} = \left| \sum_l P_{load,tot}(l,t) \cdot (phase_{distrib}(l,b) - phase_{distrib}(l,c)) \right| \tag{38}$$

$$\alpha_{unb,ca}(t)^{it=i} = \left| \sum_l P_{load,tot}(l,t) \cdot (phase_{distrib}(l,c) - phase_{distrib}(l,a)) \right| \tag{39}$$

The following equation is applied at every iteration if the VUF is over 3% at least at one load bus, and at every hour t. The index ph' represents here the different phase to phase combinations.

$$\forall t, ph' \in \{ab, bc, ca\} \ \alpha_{unb,ph'}(t)^{it=i+1} \leq \alpha_{unb,\lim}(t)^{it=i+1} \tag{40}$$

with:

$$\begin{gathered} \textit{If } VUF > 3\% \text{ and } it > 0\text{:} \\ \alpha_{unb,\lim(t)}{}^{it=i+1} = max_{ph'} \left(\alpha_{unb,ph'}(t)^{it=i} \right) \cdot \left(1 - \lambda_{unb} \cdot max_l \left(VUF(l,t)^{it=i} \right) \right) \\ \textit{Otherwise}\text{: } \alpha_{unb,\lim}(t)^{it} = \infty \end{gathered} \tag{41}$$

In these constraints, $\alpha_{unb,\lim(t)}$ is also a global limitation for the whole microgrid that only depends on the time t. Hence, Eq. (41) selects at each hour the largest previous phase-to-phase $\alpha_{unb,ph'}$ and the load bus with the largest VUF.

3.8 Objective function

The cost curve $C_{grid,buy}(t)$ tells how much it costs to buy 1 kWh from the utility grid at each hour. If the microgrid has enough resources to produce more power than its own loads require, it can either store energy in the storage system or sell it back to the utility grid. The gain from selling energy at an hour t is called $C_{grid,\,sell}(t)$. For each type of load, a parameter called value of lost load (VoLL) and denoted $C_{load}(ty)$ represents the lost gain from diminishing the use of this kind of load by 1 kW for 1 h. This parameter really depicts how valuable is each type of load. The type of load with the lowest VoLL will be the first one that the microgrid will reduce, drop out, or shift by some hours in case of emergency or high cost of electricity. On the other hand, critical loads which are important for the consumer should never be disconnected and a very high VoLL is attributed to them (e.g., important motor drives in a factory).

For the DGs, a fixed cost $C_{DG,\,on}(r)$ represents the cost of maintaining and operating a DG when this one is running. An additional variable cost $C_{DG,\,\mathrm{var}}(r)$ represents the marginal cost of producing an additional kilowatt. This last parameter is often correlated with the price of the fuel for fossil-fueled engines. The cost related to the start-up of a synchronous fossil-fueled generator is denoted $C_{DG,\,ST}(r)$.

For the storage systems, a small cost $C_{ESS,\,dis}$ is also attributed for discharging electricity from the storage equipment. This is done to prioritize the power production from renewable sources directly, rather than discharging the energy stored in the ESSs.

Since the reactive power is usually free of charges, any power electronic converter can produce or consume it. A side effect can happen in which a converter from a DER supplies reactive power to the converter of another DER. Yet, reactive power flows must be minimized to prevent voltage drops. The exchange of reactive power should indeed only be done to compensate the low power factor of certain loads or to participate actively in voltage support of the utility grid. So, to restrict the reactive power flows, a small fictive cost $C_{DER,\,react}$ is assigned to any kVAR exchanged. Furthermore, the microgrid should also prioritize its own reactive power at low voltages rather than requesting it to the utility grid at a higher voltage. The reactive power exchanged with the utility grid has been assigned with a cost $C_{grid,\,react}$, which should be higher than $C_{DER,\,react}$.

Now, the MILP objective function can be stated gathering linear and binary decision variables with their respective cost coefficients. The total cost of operation for 24 h, C_{tot}, should be minimized by tuning the various decision variables correctly and satisfying the constraints:

$$
\begin{aligned}
Min\ C_{tot} = {} & \sum_{r}\sum_{t}\left(P_{DG}(r,t)C_{DG,\,var}(r) + On_{DG}(r,t)C_{DG,\,on}(r) + ST_{DG}(r,t)C_{DG,\,ST}(r)\right) \\
& + \sum_{es}\sum_{t} ESS_{dis}(es,t)C_{ESS,\,dis}(es) \\
& + \sum_{t}\left(P_{grid,\,in}(t)C_{grid,\,buy}(t) - P_{grid,\,out}(t)C_{grid,\,sell}(t)\right) \\
& - \sum_{l}\sum_{ty}\sum_{t} P_{load}(l,ty,t)C_{load}(ty) \\
& + \sum_{es}\sum_{t}\left(Q_{ESS,\,abs}(es,t) + Q_{ESS,gen}(es,t)\right)C_{DER,\,react} \\
& + \sum_{r}\sum_{t}\left(Q_{DG,\,abs}(r,t) + Q_{DG,gen}(r,t)\right)C_{DER,\,react} \\
& + \sum_{t}\left(Q_{grid,\,out}(t) + Q_{grid,\,in}(t)\right)C_{grid,\,react}
\end{aligned} \tag{42}
$$

4. Methodology

As it can be seen in **Figure 3**, the user of the algorithm can first insert the values of different parameters in an Excel spreadsheet to create a specific scenario. Then, the algorithm first runs an initial optimization, under the form of an MILP that minimizes the daily costs, while respecting a set of constraints. The main linear variables for this optimization are either active or reactive powers. In order to quantify the impact of the PQ issues in terms of active and reactive powers, new PQ indices have been created for the purpose of this work ($\alpha_{\Delta V}$, α_{dist}, $\alpha_{unb,ab}$, $\alpha_{unb,bc}$, and $\alpha_{unb,ca}$). These new indices have the advantage to be directly computed in the MILP and do not require any load flow (see Section 3.7.2). The MILP optimizations are solved with the software GAMS. The solver typically converges in around 90 iterations within a relative error of 2×10^{-6}.

During this first optimization, the parameters $\alpha_{\lim,\Delta V}$, $\alpha_{\lim,dist}$, and $\alpha_{\lim,unb}$ have been initialized to a very large value, high enough so that the inequalities (32), (35), and (40) do not restrict the value of $\alpha_{\Delta V}$, α_{dist}, $\alpha_{unb,ab}$, $\alpha_{unb,bc}$, and $\alpha_{unb,ca}$.

Using the results of the MILP, the program OpenDSS [19] will launch a set of harmonic load flows. These load-flows will give all the information about the voltage on the different nodes of the microgrid that is needed to compute the PQ indices, i.e., the voltage deviation in per unit, the voltage THD, and the VUF. If any of these indices does not respect the standards that have been presented in the Section 2, a noninfinite value will be attributed to the corresponding α_{lim}, so that at

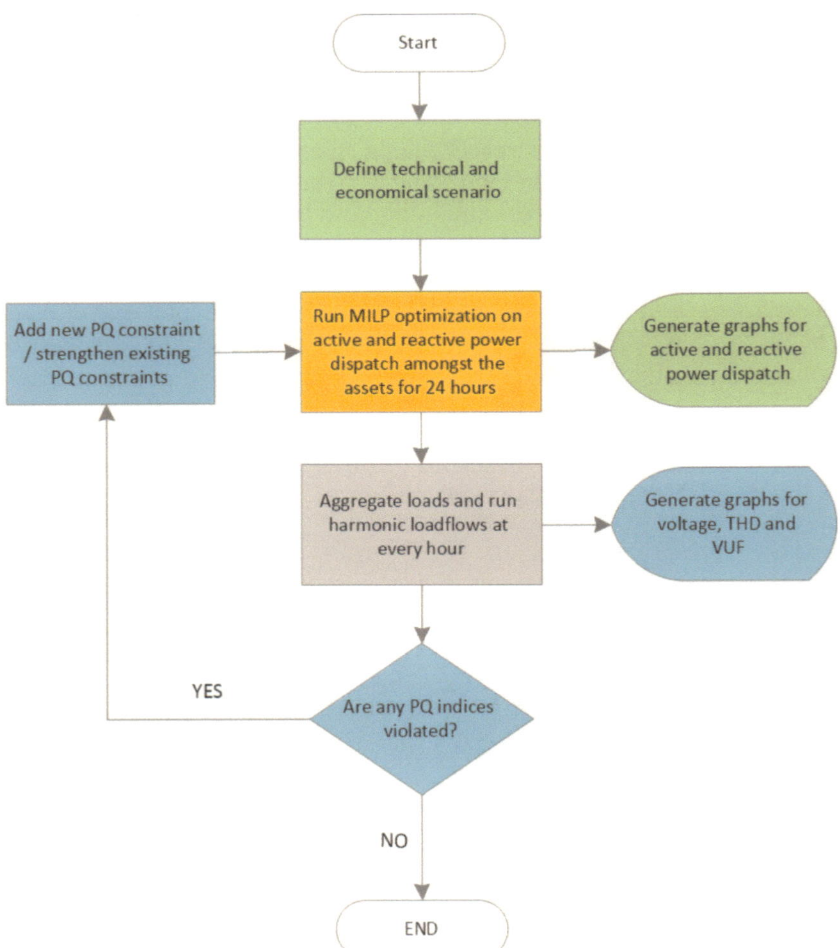

Figure 3.
Flow chart of the decision process. Green processes are performed by Excel, orange by GAMS, gray by OpenDSS, and blue by MATLAB.

least one of the PQ constraints shown in (32), (35), and (40) is truly restrictive for the next MILP optimization.

5. Simulation for different case studies

The conceptual microgrid that has been used in this chapter is shown in **Figure 4**. It is an alteration of the one presented in [6]. While the properties of the lines, the American voltage levels, and the characteristics of the transformers and the DERs have been kept identical, this microgrid has different kinds of loads and a smaller number of branches. In order to show the diversity of power consumers in a microgrid, three main different loads have been introduced: a residential load, an industrial load, and a commercial load.

The test microgrid has three different levels of voltage, at 11.2 kV, 408, and 207 V. They are interfaced by transformers but do not include any DC bus. The architecture is radial, as usual in distribution systems, and divided into four different branches. The first branch contains two diesel generators $genset_1$ and $genset_2$. These diesel generators are essentially a backup power supply in stand-alone operations. The second branch has a first bus at 408 V (node 3), which is connected to a small 60 kWp wind turbine and a battery storage system of 80 kWh useful energy. This branch continues to the 207 V voltage to feed a residential load of 48 dwellings, called $Load_1$. The third branch supplies an important industrial load named $Load_2$. Finally, the fourth branch is a bidirectional line that reaches a small office building load, $Load_3$, with a 40 kWp PV installation. The power demands of the different aggregated loads are represented in **Figure 5**.

The utility grid seen from the PCC at node 4 has been replaced by its Thevenin equivalent. It has a short-circuit power of 1000 MVA and a X/R ratio of 22. All

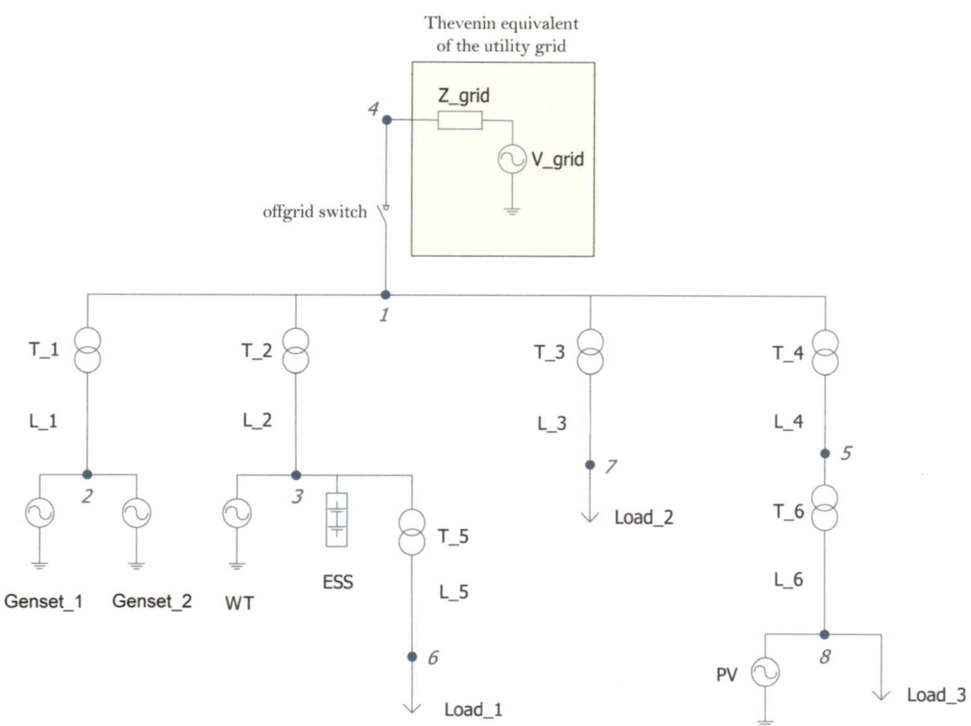

Figure 4.
Architecture of the test microgrid.

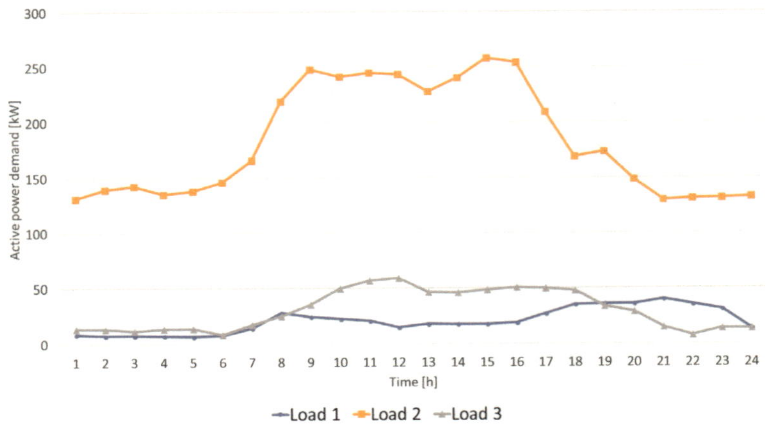

Figure 5.
Active power demand curves of the three aggregated loads of the conceptual microgrid.

the technical and economical parameters concerning the properties of the lines, the transformers, and the DERs are listed in Appendix. The loss parameters $loss_P$ and $loss_Q$ are tuned to the upper bound of the ratio between the total active (or reactive) loss with respect to the active (or reactive) total load. After several trials, it has been found that $loss_P$ is worth 1.5% and $loss_Q$ is worth 13.0%.

The VoLL and the flexibility of each type of load for this test-case microgrid are reported in **Table 1**, alongside other characteristics from [20–24].

5.1 Normal operation

The first investigated scenario aims to represent the operation of the microgrid under standard technical and economic conditions. In order to establish a realistic price curve for the utility grid electricity throughout the day, the data are directly taken from the hourly values of the day-ahead prices from the Belgian power exchange (BELPEX) on March 2, 2018 [25]. The tariffs for the system operators and regulators and the other public obligations reproduce the one imposed by the Brussels DSO Sibelga on middle-voltage clients [26]. **Figure 6** shows the composition of the cost of electricity from the utility grid, translated into dollars with a fixed exchange rate of 1.1942$/€. The reactive power flow required by the grid at each hour for the voltage support has been fixed arbitrarily, but it remains small with respect to the reactive power from the load and does not impact substantially the results.

Type of load	C_{load} (VoLL) ($/kWh)	*Flex* (%)	THD_I (%)	*pf*	$Type_{ratio}$		
					$Load_1$ (%)	$Load_2$ (%)	$Load_3$ (%)
HVAC	2	20	8.2	0.98	23.9	16.8	37.3
DHW	3	20	0	1	9.5	14.4	1.8
Lights	5	10	27.1	0.8	9.4	6.5	10.6
Appliances	5	35	43.3	0.65	57.2	7.3	50.3
Motor drives	10	1	0%	1	0	55	0

Table 1.
Properties of the types of loads.

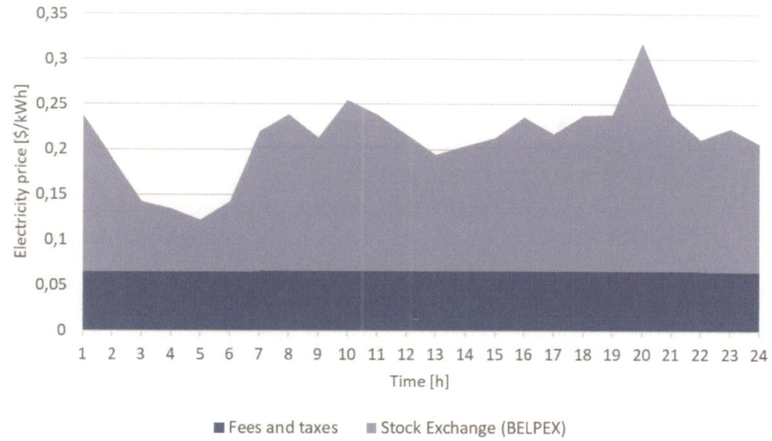

Figure 6.
Composition of the price of electricity from the utility grid for MV clients on the March 2, 2018 in the Brussels-Capital region.

The *Normal operation* scenario depicts how the microgrid operates without PQ issues, when the utility grid is reliable and its cost profile is based on real historical data. **Figure 7** shows the results of the simulation concerning the active power operation points of all the DERs and the electricity drawn from the utility grid to cover the total load. The active power losses are not represented in this graph because they are relatively small, usually below 3 kW, and can be neglected. It can be seen that the diesel generators never appear in the electricity mix in this scenario. This is due to the fact that the different costs associated with their operation are quite high compared to the cost of electricity bought on the spot market. The renewable sources are obviously always running because their variable cost is extremely low. Concerning the ESS, it is charged when the grid is the lowest, at 3, 4, 5, and 13 h and discharged three times, at 1, 10, and 20 h, when the price of electricity peaks (**Figure 6**).

Figure 8 shows the voltages in per unit at the important nodes of the network. The 5% deviation from 1 pu is never exceeded during 24 h. **Figure 9** represents the voltage THD evolution at each load node throughout the day. As one can observe, the $Load_1$ and $Load_3$ have a low-voltage THD, below 1.5%. For the industrial $Load_2$, it is practically negligible because of the scarcity of nonlinear devices. Since all the

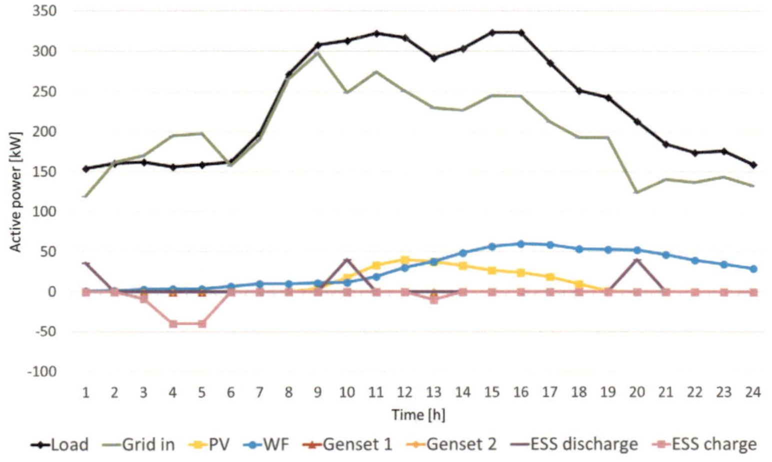

Figure 7.
Active power dispatch during the day for the "normal operation" scenario.

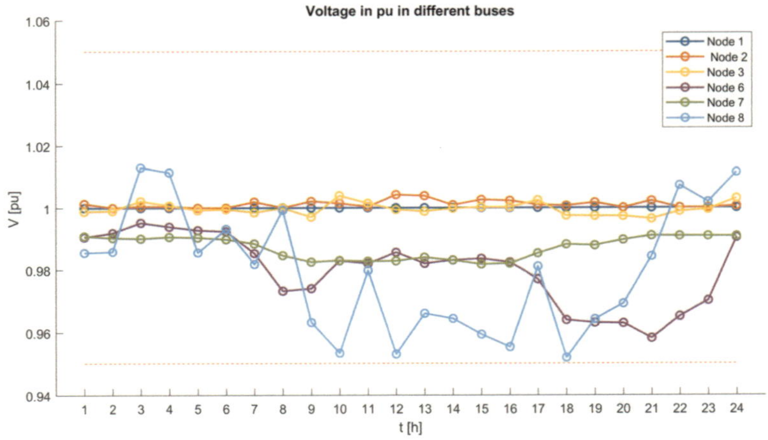

Figure 8.
Evolution of the voltage in per unit for the scenario "normal operation."

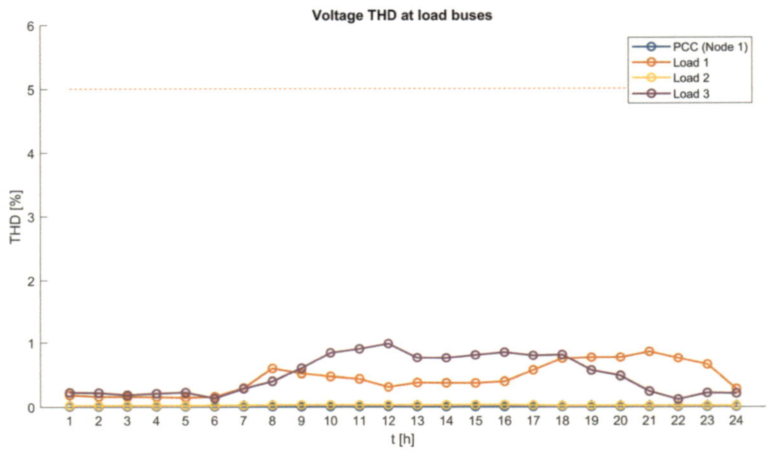

Figure 9.
Evolution of the voltage THD in percent for the scenario "normal operation."

aggregated loads are assumed perfectly balanced in this scenario, the VUF is equal to 0 at every node.

Since none of the PQ norms are violated in this scenario, the constraints in (32), (35), and (40) are not activated and there is no need for iterations between the MILP and the harmonic load flows.

5.2 Voltage drop issue

To illustrate the action of the EMS algorithm when a power quality issue is detected, a scenario is created concerning a potential voltage drop. Other scenarios on harmonic distortion and phase unbalance have been conducted in a longer version of this study and essentially show similar results in terms of accuracy and speed of convergence.

In this scenario, the load curves of $Load_1$ and $Load_3$ are increased by 10% in comparison with the **Table 1**. This scenario was created to test the resilience of the microgrid toward voltage deviations. After the initial optimization, the EMS detects with the load flows that the increase of the total power requested by the first load causes several voltage drops under the 0.95 bound, at 10 h, from 12 till 16 h, and at

18 h. This first observation can be seen in **Figure 10**. The voltage at the node 8, where the $Load_3$ is connected, is only worth 0.9307 in per units at 12 h. To address this problem, an iteration process is started and activates the constraint (31), with a parameter $\lambda_{\Delta V}$ equal to 0.9. The value of this sensitivity parameter has been found by a trial-and-error process. It comes from a trade-off between the speed and the accuracy of the convergence. The next graph on **Figure 11** shows the reduction of the voltage drop after the third iteration. It can be concluded that the EMS manages to keep the voltage above 0.95 pu after only two iterations. The voltage obtained at 12 h after the first iteration is 0.9350 and then 0.9551 after the second iteration. Still for this particular hour, the DSM scheme has reduced the HVAC consumption in $Load_3$ by 20% and the appliance share by 11.41%. The appliances have indeed a low fundamental-frequency power factor, which leads to a higher reactive power demand and a worse effect on the voltage drop.

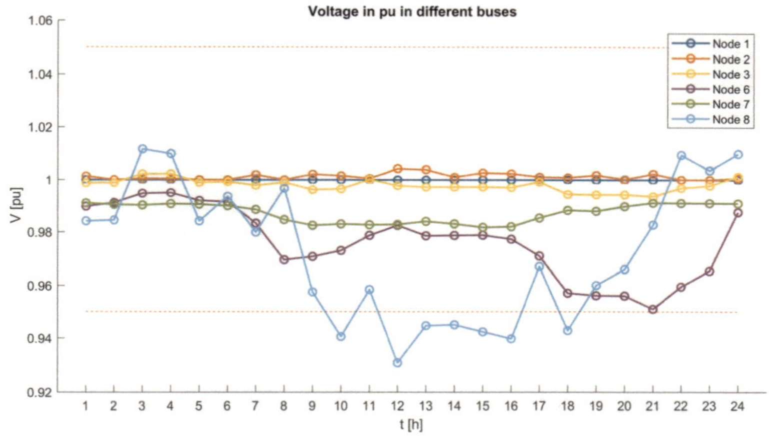

Figure 10.
Voltage in per unit at the different nodes for the initial stage.

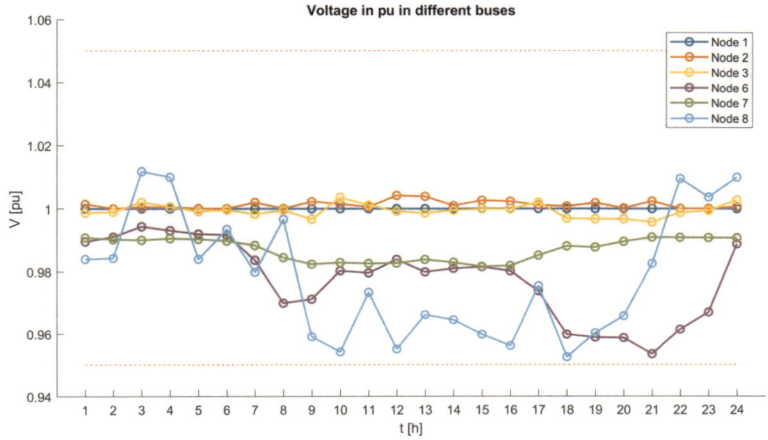

Figure 11.
Voltage magnitude in per unit after the first optimization for the different nodes after two iterations of the EMS algorithm regulation loop.

6. Conclusion and perspectives

Security of electricity supply, flexibility, cost-effectiveness, and renewable sources integration are some of the motivations that lead private or public entities to consider the implementation of grid-connected microgrid. The local production and storage systems allow some microgrids to ensure the supply of electricity for their loads in case of a frequency or voltage outage on the traditional grid. However, studies have shown that PQ disturbances can be difficult to tackle in small-scale microgrids, due to the lower stiffness of the distributed power generation.

The purpose of this chapter is to tackle several of these PQ issues that occur in steady state, namely voltage drop, harmonic distortion, and phase unbalance, by acting on the demand level of certain types of electrical devices in the microgrid. First, it has been considered that any load at a node of the network can be represented as the aggregation of several smaller loads that account for a particular type of end use. Consequently, every type of load has its own physical and economical properties. A demand-side management framework has been implemented, so the demand level can be regulated within a certain flexibility range. The DSM mechanism can be launched whenever the cost of electricity peaks, in offgrid situations, or to mitigate a PQ issue that violates the standards. This process is integrated under the form of constraints inside an optimization-based EMS algorithm that minimizes the overall daily costs. The algorithmic structure of this tool consists in a regulation loop between an MILP optimization and harmonic load flows. In order to reflect the magnitude of the PQ issues in the MILP optimization, new power quality indices that rely on active and reactive powers have been introduced. The designed algorithm has been simulated according to different scenarios on a test-case microgrid. The results show that the production, the storage, and the consumption in the microgrid can adapt efficiently to the price of electricity from the traditional grid and that the different power quality standards can be met after few iterations. This algorithm can be used for grid-connected microgrids at low or medium voltage that possess an efficient communication framework between the different consumers and producers of electricity.

Further research on the topic of power-quality-supporting EMS algorithms could include the possibility to add electric vehicles (EV) management, combined heat and power (CHP) generation units, or other kinds of electrical equipment inside the microgrid.

A. Appendix

Tables 2–7 gather the data used to run the simulations on the test-case microgrid presented in the fifth section.

Transformers	Voltage (kV)	Base MVA	Connections	%Z	X/R
T1	0.48/11.2	2	Δ-Y	5.75	6
T2, T3, T4	11.2/0.48	0.5	Δ-Y	5.75	6
T5, T6	0.48/0.207	0.25	Δ-Y	5.75	3

Table 2.
Properties of the transformers.

Lines	Voltage (kV)	Type	r ($\Omega\ km^{-1}$)	x ($\Omega\ km^{-1}$)
L1, L2, L3, L4	0.48	3 phase/4 wire	0.049	0.027
L5, L6	0.207	3 phase/4 wire	0.06	0.03

Table 3.
Properties of the lines.

DG	Voltage (kV)	P_{min} (kW)	P_{max} (kW)	$Q_{abs,\ max}$ (kVAR)	$Q_{gen,\ max}$ (kVAR)
genset1, genset2	0.48	50	200	150	150
WT	0.48	0	60	45	45
PV	0.207	0	40	30	30

Table 4.
Technical properties of the DGs.

DG	$C_{DG,\ on}$ (\$/h)	$C_{DG,\ \text{var}}$ (\$/kWh)	$C_{DG,\ ST}$ (\$)
genset1	100	0.90	300
genset2	100	0.91	300
PV	0	0.01	0
WF	0	0.01	0

Table 5.
Economic properties of the DGs.

ESS	Voltage	$P_{char/dis,\ max}$	$Q_{abs/gen,\ max}$	Nominal energy	Init SOC	Roundtrip efficiency
ESS1	0.48 kV	40 kW	40 kVAR	80 kWh	50%	81%

Table 6.
Technical properties of the ESS.

Harmonic order	Current harmonic magnitude (%fund.)				
	HVAC (%)	DHW (%)	Lights (%)	Appliances (%)	Motor drives (%)
3	5.0	0	21.1	29.9	0
5	6.0	0	11.9	23.3	0
7	2.3	0	11.8	15.7	0
9	1	0	2.0	10.8	0
11	0	0	1.0	8.2	0
THD_I	8.2	0	27.1	43.3	0

Table 7.
Current harmonic spectrum of the types of load.

Author details

Gaspard d'Hoop[1], Olivier Deblecker[1*] and Dimitrios Thomas[2]

1 Electrical Power Engineering Unit, Faculty of Engineering, University of Mons, Mons, Belgium

2 Faculty of Engineering and ERA Chair 'Net Zero Energy Efficiency on City Districts', Research Institute for Energy, University of Mons, Mons, Belgium

*Address all correspondence to: olivier.deblecker@umons.ac.be

References

[1] Lavoine O. Thoughts on an Electricity System and Grid Paradigm Shift in Response to the EU Energy Transition and the Clean Energy Package. Florence School of Regulation Advisory Council; 2018. pp. 1-5. DOI: 10.2870/535108

[2] FutureReady. Europe Experiments with Transparent Pricing—Is It a Viable Option in the U.S.? [Internet]. 2016, *Landis+Gyr*, 10 2016. Available from: https://www.befutureready.com/ezine-article/europe-experiments-transparent-pricing-viable-option-u-s/ [Accessed: May 3, 2018]

[3] Ali A et al. Overview of current microgrid policies, incentives and barriers in the European Union, United States and China. Sustainability. 2017;**9**: 1-28. DOI: 10.3390/su9071146

[4] Hossain E et al. A comprehensive study on microgrid technology. International Journal of Renewable Energy Research. 2014;**4**:1094-1102. ISSN: 1309-0127

[5] Khalid S, Dwivedi B. Power quality: An important aspect. International Journal of Engineering, Science and Technology. 2010;**2**:6485-6490. ISSN: 0975-5462

[6] Hong M et al. An energy scheduling algorithm supporting power quality management in commercial building microgrids. IEEE Transactions on Smart Grid. 2016;**7**:1044-1056. DOI: 10.1109/TSG.2014.2379582

[7] Zhao Z et al. Power quality improvement with SVC in power supply system. In: Proceeding of the 2012 China International Conference on Electricity Distribution (CICED '12); September 10–14, 2012; Shanghai. New York: IEEE; 2012. pp. 1-4

[8] Jin X et al. Hierarchical microgrid energy management in an office building. Applied Energy. 2017;**208**: 480-494. DOI: 10.1016/j.apenergy.2017.10.002

[9] Chitra N et al. Survey on microgrid: Power quality improvement techniques. ISRN Renewable Energy. 2014;**2014**:7

[10] Chattopadhyay S, Mitra M, Sengupta S. Electric Power Quality. 1st ed. Dordrecht: Springer Netherlands; 2011. 182p. DOI: 10.1007/978-94-007-0635-4

[11] Andrei H et al. Fundamentals of Reactive Power in AC Power Systems. Bucharest: Springer International Publishing; 2017. 248p. DOI: 10.1007/978-3-319-51118-4

[12] Institute of Electrical and Electronics Engineers (IEEE). IEEE Guide for Design, Operation, and Integration of Distributed Resource Island Systems with Electric Power Systems. IEEE Std 1547. Vol. 42011. pp. 1-54. DOI: 10.1109/IEEESTD.2011.5960751

[13] Share Pasand MM. Harmonic aggregation techniques. Journal of Electrical and Electronic Engineering. 2015;**3**:117-120. DOI: 10.11648/j.jeee.20150305.13

[14] Mazin HE, Xu W. Harmonic cancellation characteristics of specially connected transformers. Electric Power Systems Research. 2009;**79**:1689-1697. DOI: 10.1016/j.epsr.2009.07.006

[15] Testing and measurement techniques – General guide on harmonics and interharmonics measurements and instrumentation, for power supply systems and equipment connected thereto. International Electrotechnical Commission. IEC

61000-4-7. IEC International Standards. 2002;**2**:1-71

[16] Grady WM, Santoso S. Understanding power system harmonics. IEEE Power Engineering Review. 2001;**21**:8-11. DOI: 10.1109/MPER.2001.961997

[17] Mumtaz F, Bayram IS. Planning, operation, and protection of microgrids: An overview. Energy Procedia. 2017;**107**:94-100. DOI: 10.1016/j.egypro.2016.12.137

[18] Thomas D, Deblecker O, Ioakeimides C. Optimal operation of an energy management system for a grid-connected smart building considering photovoltaics uncertainty and stochastic electric vehicles driving schedule. In: Proceeding of the 43rd Annual Conference of the IEEE Industrial Electronics Society (IECON '17); 29 October–1 November 2017; Beijing. New York: IEEE; 2017. pp. 3621-3626

[19] Electric Research Power Institute. Simulation Tool—OpenDSS [Internet]. 2018. Available from: http://smartgrid.epri.com/SimulationTool.aspx [Accessed: March 2, 2018]

[20] Pipattanasomporn M et al. Load profiles of selected major household appliances and their demand response opportunities. IEEE Transactions on Smart Grid. 2014;**5**:742-750. DOI: 10.1109/TSG.2013.2268664

[21] Majithia CA, Desai AV, Panchal AK. Harmonic analysis of some light sources used for domestic lighting. Lighting Research and Technology. 2011;**43**:371-380. DOI: 10.1177/1477153510394597

[22] Today in Energy (U.S. Energy Information Administration). American Households Use a Variety of Lightbulbs as CFL and LED Adoption Increases [Internet]. 2017. Available from: https://www.eia.gov/todayinenergy/detail.php?id=31112 [Accessed: March 7, 2018]

[23] Nikum K, Saxena R, Wagh A. Harmonic analysis of residential load based on power quality. In: Proceeding of the IEEE 7th Power India International Conference (PIICON '16); November 25–27, 2016; Bikaner. New York: IEEE; 2016. pp. 1-6

[24] United States Energy Information Administration (EIA). Annual Energy Outlook 2018 [Internet]. 2018. Available from: https://www.eia.gov/outlooks/aeo [Accessed: June 6, 2018]

[25] ENTSOE-E Transparency Platform. Day-Ahead Prices in Belgium [Internet]. 2018. Available from: https://transparency.entsoe.eu/transmission-domain/r2/dayAheadPrices/show [Accessed: April 24, 2018]

[26] Sibelga. Obligations de service public (OSP) [Internet]. 2018. Available from: https://www.sibelga.be/fr/tarifs/tarifs-utilisation-reseau/obligations-de-service-public/osp-electricite [Accessed: April 25, 2018]

Section 3

Protection

Chapter 5

Microgrid Protection Systems

Mylavarapu Ramamoorty and
Suraparaju Venkata Naga Lakshmi Lalitha

Abstract

Micro grids are miniature version of conventional large power grids functioning either autonomously or with inter connection to the main grid. Primary function of micro grid is to serve power at distribution level. Distributed energy resources (DERs) connected to the micro grid enables reliable and efficient operation of micro grid. Protection of micro grids assumed importance due to increased penetration of distributed energy resources. Most of the distribution systems in earlier days are radial in nature and protection systems are designed for that. These protection systems pose serious challenges when applied to present day distribution systems which are mesh connected and fed by the distributed energy resources. Limitation of the conventional protection scheme demands new insights and methodologies for micro grid protection. Due to intermediate current injection from DERs the conventional coordination of over current (O/C) relays is not possible. Further in meshed systems the fault current flow is bidirectional. Hence the protection of micro grid systems with DERs require different approach to ensure faults are cleared in less time and minimal number of consumers connected to the system are affected. A comprehensive analysis of the suitable techniques applicable for micro grid protection is presented in this chapter.

Keywords: renewable energy sources, distributed energy resources, micro grid, distribution systems, protection, over current relay, distance relay, differential relay

1. Introduction

Protection is a vital aspect of power system which needs lot of attention everywhere. Majority of the existing protection techniques for distribution systems are developed for radial distribution lines. These techniques will not be directly applicable to the micro grids with meshed network in the presence of distributed energy resources (DER). A CIGRE definition of micro grid is given as *Microgrids are electricity distribution systems containing loads and distributed energy resources, (such as distributed generators, storage devices, or controllable loads) that can be operated in a controlled, coordinated way either while connected to the main power network or while islanded* [1]. A typical CIGRE benchmark LV micro grid is shown in **Figure 1** [2].

The role of DERs in the present and future distribution systems is inevitable. Deployment of distributed generators (DGs) proved to be very effective means of meeting the ever increasing energy needs and concerns for Environment pollution and the depletion of fossil fuels. Employability of proper protection schemes to suit the micro grid environment fed by the renewable energy resources has assumed lot of importance. Protection of micro grids poses several challenges for the utility

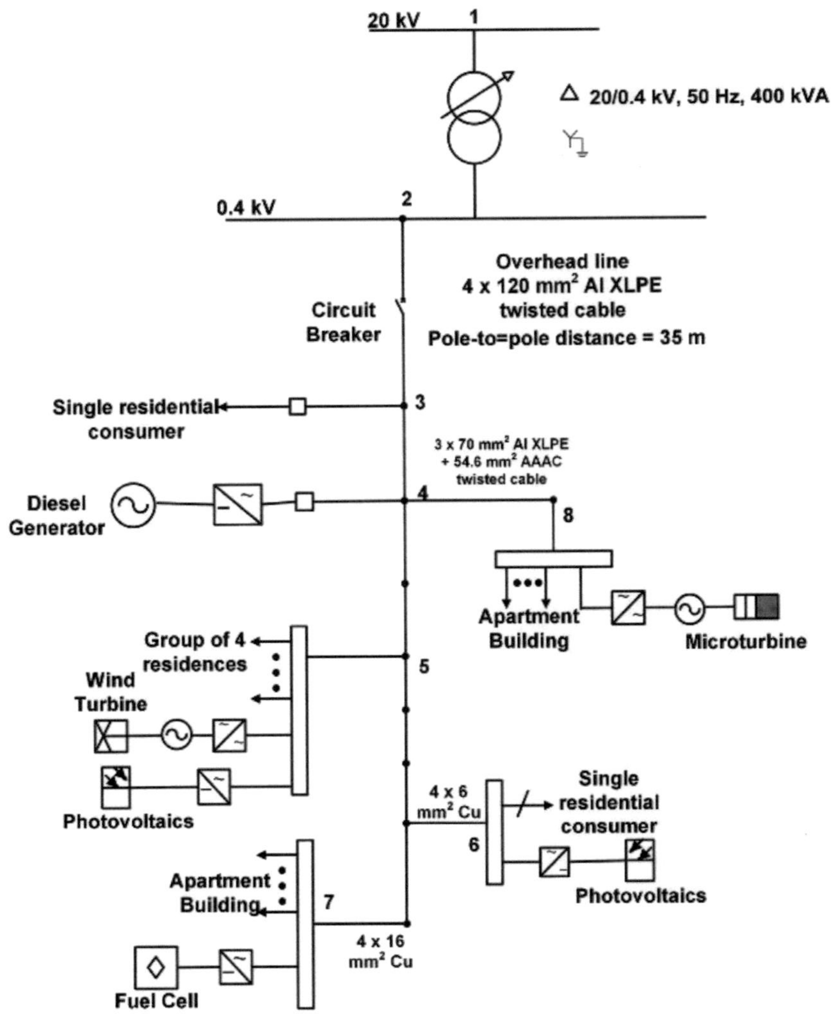

Figure 1.
CIGRE LV benchmark microgrid.

engineers. Protection of micro grids opened the doors for various investigations by the researchers across the globe. Some important aspects related to the protection issues of micro grids are presented in this article.

General protection methods applied to the distribution network are designed for radial systems having unidirectional power flow. With DGs power flow is no longer unidirectional and it causes a serious threat when conventional protection methods are used for the micro grid with DGs. Another concern is that the micro grid is expected to operate safely in grid connected or islanded mode. The intermittent nature of the output power from a DG makes the selection of the operating characteristics of the relays to be complicated. Further, most of the DGs are connected to the grid through converters which have independent control strategies. Limited fault current of the inverter based DGs and maintenance of Fault ride through capability should be given due consideration in protection. Locating the fault and proper isolation of the fault are also important [3, 4].

Time graded and current graded over current protective schemes have been in use for the radial distribution systems. Distance and differential protection schemes are also employed. Voltage based protection and THD (total harmonic distortion) based protection are found to be suitable for protection of micro grids with DGs.

Adaptability is the need of any protection method used for micro grid. The continuous change in the network configuration due to the addition of DGs and/or future expansions necessitates that the protection equipment must be adaptable as per the requirement [5]. Communication is another aspect of the protection of the micro grid. IEDs (intelligent electronic devices) are being used for this purpose in the grid. Suitable communication protocols are developed and IEC 61850 is being followed. Protection plays a key role in the development of micro grids due to the increase in the number of DGs, IEDs, storage systems and the requirement of a suitable communication medium [6].

This review article covers the need for changes to be made to the conventional protection systems when applied to micro grids in general and discusses recent advances made in the field of micro grid protection. Brief and critical review of the recent papers published on this subject is included. It is expected that this review article will provide a bird's eye view of the status of protection systems adapted for the micro grids with DERs.

This article comprises of six sections with the introduction as first section. Section 2 discusses conventional distribution system protection along with the deficiencies of the conventional O/C protection systems as applied to typical micro grids. Third section deals with brief description of the renewable energy sources (RES) and the need to replace the conventional generation systems considering environmental considerations. Configuration of micro grids with DERs is explained in Section 4. Problems of interfacing micro grids with conventional grid will be discussed in this section. Section 5 presents critical review of the recent papers dealing with the protection of micro grids. Section 6 concludes the article.

2. Conventional distribution system protection

Any protection system must be simple, fast, reliable and consistent apart from being selective and sensitive to the faults. Any protection system should not operate under normal conditions and must operate under abnormal conditions ensuring security and dependability of the protective system. These are the two important reliability indices which need to be optimized always. The two main classes are the radial distribution system and the meshed system.

For a radial feeder, fault current flows in one direction only as there is a single source of power. Relay setting in this case is relatively easy. This makes designing of strategies for protection become very straightforward for distribution systems typically. Simple devices such as reclosers, fuses and over current relays are used for protection. As a thumb rule fuses are set to operate for permanent fault and reclosers are set for temporary fault clearance. This is done as a part of fuse to recloser coordination with the intention of saving the fuse and also allow for the temporary faults to clear themselves with fast recloser action. Fuse to fuse coordination, relay to relay coordination and relay to fuse coordination are also required to be done. This is to ensure that minimum number of consumers connected to the distribution system are affected. Generally the fuse to fuse coordination is done from characteristic curves or selectivity tables supplied by manufacturer. In relay to relay coordination, time graded/current graded/combination of time and current grading is employed. Definite time, inverse time O/C relays are used. Inverse definite minimum time relays allow the protection engineer for flexible settings of the relay. Discrimination time of 0.5–0.3 s is possible with the fast acting relays and circuit breakers. In relay to fuse coordination, time margin is computed by taking into consideration, the operating time of the upper fuse for proper relay setting. It is essential that for proper coordination, fault current flowing through the protective devices must be between the set minimum and maximum fault current that is possible and the fault current through all protective devices are almost equal. It is important to note that in case of a radial feeder, ensuring continuity of supply to maximum possible number of consumers after clearance of sustained fault is not possible. Ring main distribution system is an alternative [7].

In ring main system, each load can be supplied power from two different paths. In case of a fault in one feeder, the other feeder continues to supply whole or a percentage of total loads. Directional O/C relays are used along with non-directional O/C relays to minimize the number of consumers affected. Grading of the directional O/C relays starts from the load end to the source in the ascending order of the time, whereas for the non-directional O/C relays time discrimination is from the source side to the load side. In case the ring main is supplied by more than one source, coordination among the relays is not that easy. If two sources are present, the ring is opened at one end usually at one of the sources and the grading is done by presuming the other source as a single source. Employing differential protection for the section between the two sources is another practice. In this case, the rest of the system is treated as being fed from a single source. If more than two sources are present, then the design of protection system becomes more involved [8].

Most of the conventional protective systems are designed for the radial distribution systems where there is only one directional power flow. Design of a proper protection system that can be adapted for radial or ring main system with more than two sources has challenges posed due to bidirectional power flows that are encountered. Dependability of the usage of conventional protection scheme which is suitable only for radial system is very low and is therefore not recommended for the modern meshed distribution systems with DERs. It would be economical if the existing protection system can be modified or upgraded to match the protection needs of the modern system rather than discarding the old systems and going for an altogether new protection system. It is very expensive and is not advisable. Ways and means of using the existing protection system without losing the important aspects of protection is highly desirable [9]. Lot of research is focused in this direction to find out effective utilization of the existing infrastructural facilities of the convention distribution protection systems.

3. Need of renewable energy sources (RES)

In order to meet the increased demand and to reduce the transmission/distribution losses, generation at the load points is being done using the RES. There is a pressing need to redesign the conventional protection systems incorporating the RES. Another point worth mentioning is that the conventional protection systems are designed for radial systems expecting large values of fault currents. With the introduction of the RES, there are two possible modes of operation namely grid connected mode and islanded mode. In case of grid connected operation, there is a possibility of large magnitude fault currents but may not be always true in islanded operation and it poses serious concerns related to protection. Inclusion of a DER will causes bidirectional power flows in the distribution system. It reduces the possible upper and lower limits of fault current along with reduction in the fault current through protective devices posing serious threat to the conventional protection coordination [10].

DERs and their associated control, communication and protection devices have become an integral part of modern distribution systems. In any distribution system, if the penetration of these DERs is more in any area then that geographical area is being referred as a micro grid. Micro grid is a part of the main distribution grid and it operates independently to some extent. Major element of a micro grid is a

DER. DER can be a PV cell, fuel cell, wind turbine, diesel generator, energy storage system mainly based on Battery, etc. There are many advantages of DERs like reduction of transmission and distribution losses, eco-friendly power generation reducing the carbon emission, possible reduction in congestion in the networks, enhance the energy efficiency by proper utilization of the solar and wind energy.

Major differences between a conventional distribution system and micro grid can be categorized into three parts namely interfacing of the inverter fed DGs, grid connected and islanded operation and bi directional power flows. A micro grid is expected to operate successfully and independently even when there is a disturbance in the main grid. The main challenges posed in the protection of micro grids when compared to conventional system are listed below.

1. Sensitivity and selectivity of the protective relays gets affected due to the local generation by the DG. Settings should be done in such a way that protection is ensured even in islanded mode of operation.

2. Due to the presence of DGs interfaced to the grid through inverters, fault current seen by the relays is reduced during islanded operation. It affects the protective action by the relays in terms of, either delay in the protective action or non-detection of the fault.

3. DGs will also affect the maximum and minimum fault currents through a feeder and it results in serious coordination problems between recloser, fuse and O/C relays.

4. Undesired tripping of the non-directional O/C relay in a healthy feeder when the DG feeds a fault outside the healthy feeder. It happens because the DG tries to feed the fault through the healthy feeder.

5. An auto recloser clears a temporary fault by fast opening and reclosing of the circuit. If a synchronous DG is present in the micro grid, during this auto recloser action, it might experience a slight shift in synchronism. In that case, recloser will be connecting two systems which are not synchronous causing a serious threat to the entire system. Also the DG will be trying to maintain the system voltage and in turn the arc at the fault location. It might make the temporary fault to appear as a permanent fault.

4. Micro grid and DER integration

Integration of DER is an important aspect of micro grid operation. There are different control strategies applied in micro grid operation. Basically these can be classified as overall network control and DER control. Supervisory control of the network is done in centralized and decentralized mode using distribution management systems (DMS). DER control is normally chosen depending on the circumstances considering the network operation scenarios and the interaction with other DERs. In grid connected mode real and reactive power control is adopted where as in islanded mode frequency and voltage control is used [11].

In a micro grid, apart from DER, there are many other types of equipment such as data interfaces, monitoring devices, communication protocols, protective devices etc. Communication is another important element of modern distribution systems. Effective communication protocols have been established and standardized for use in substations. IEC 61850 is a global standard communication protocol which plays a significant role in all aspects of distribution system viz. control, metering and protection. If the state of the micro grid is subjected to frequent changes due to intermittent nature of DGs and changes in load profile, operation strategies of different equipment need to be adjusted accordingly. Thus the system integration efficiency depends on the equipment integration. Further, the conversion of the operating mode of the micro grid from grid connected mode to islanded mode or vice versa also demands the adjustments in operation strategies of different equipment. IEC 61850 provides a flexible architecture, service and service essential for interoperability and upgrading required for various needs of modern distribution systems.

IEDs (intelligent electronic devices) are required as the devices are expected to be intelligent enough for data acquisition, transmission to control centres as well as decision making whenever necessary. These devices are being used extensively and are having the latest technology for sensing. It allows for two way communication and greater awareness on the situation in the power distribution system. These devices can be controlled remotely thus allowing efficient operation during disturbances. Another feature of the IEDs is that they can communicate with other devices present in the system allowing effective fault identification and restoration. With the application of FPGA technology, IEDs are becoming more effective [12].

As the micro grid is interconnected to the main grid, it is essential that the protective system must ensure the safety for faults in micro grid as well as for the faults in main grid. In case of a fault in main grid, micro grid should be isolated such that the consumers supplied by micro grid are not affected. If the fault is in the micro grid itself, then smallest possible percentage of consumers must be disconnected. Under these two circumstances, many challenges are there in the protective system design [13]. Some points to be considered while designing the protective system are (i) intermittent nature of the power generation by DGs due to changes in solar power, wind power, etc., (ii) variations in the load (iii) number of DGs, (iv) type of DGs such as inverter fed DG or synchronous DG, etc., and (v) topology of the network. In the grid connected mode, islanding may result accidentally or incidentally due to faults/human error/intentional opening for servicing/faulty operation of protective devices/natural disasters/and equipment failure. IEDs are employed for control and protection in modern distribution systems. Active management of the network and adaptive protection is possible through IEDs [14]. Inverter based DERs are expected not to get disconnected following a fault or contingency immediately. They should possess the ability to remain connected to the Grid for some time. It is called Fault ride through (FRT) capability. It is necessary to have sufficient fault current for the relays to sense the fault and to maintain the voltage during any contingency. Unlike a synchronously connected DER, inverter based DERs do not possess the FRT capability inherently [3]. FRT requirements in micro grids can be easily accomplished with IEDs by employing suitable controllers for inverters. To change over the protection strategies when the micro grid isolates from the main grid either intentionally or otherwise there is a need to detect quickly such isolation and secure the micro grid. The detection techniques adapted for sensing isolation and taking appropriate action for the controls and protection are outlined in the next section.

5. Islanding detection and recommended practices for micro grid protection

An efficient protection scheme must ensure proper protection to the micro grid in its both modes of operation, i.e., grid connected mode and islanded mode. It also should ensure proper functionality during the transition from one mode to another depending on the requirement. The topological network changes due to the transition from one mode to other demands for the changes in the settings of the protective relays. Before proceeding further, one must understand the nature of the fault currents during grid connected mode and islanded mode. There are several factors that need to be taken into consideration such as the size of the DERs, type of DER, no. of DERs, how they are integrated to the main grid and the islanding detection methodologies. Many functional differences in the operation of a synchronous DER and inverter based DER calls for alternative protection strategies for them. Initially the effect of micro grid operation on the fault currents is discussed in this section. Later, general categories of O/C protection, distance protection, differential protection along with voltage based methods applied to suit the requirements of micro grid in grid connected and islanded mode are discussed. Adaptive protection is the main ingredient of micro grid protection.

5.1 Effect of micro grid integration with main grid

Micro grid is integrated with the main grid with an interfacing switch. As per the IEEE standard 1547-2003, a DG should be immediately get disconnected for any type of fault occurrence in the grid [15]. If a fault occurs anywhere in the main grid or micro grid, the static switch connecting the two gets opened and thus the micro grid goes into the islanded mode of operation. Along with opening the static switch, location of the fault also should be detected simultaneously. If incase, the faults happens to be within the micro grid, a suitable protective system should be brought into operation immediately. It should be done online, i.e., detection of the fault location and the initiation of protective action within the micro grid [16]. For faults in the main grid the static switch opens and islands the micro grid so that the DGs do not contribute to the fault current. Then the control system for islanded operation comes into play. For faults in the micro grid the static switch opens to remove the fault current contribution from main grid and the protection system of the micro grid comes into play and clears the fault. Therefore, one should recognize the importance of islanding detection preceding the protective action. In time and accurate detection of islanding is essential for fulfilling adequate protection requirements in micro grid operation.

Few points worth mentioning are as follows:

1. In inverter based DER, fault current gets affected by the limitation of 2 pu rated current of the interfacing inverter [17].

2. If the DER is connected to a single phase load, it might result in considerable unbalance between the phases in a three phase system.

3. Intermittent nature of the power output of a DER throws serious challenges in assessing the possible fault currents and relay settings.

4. Short circuit level of the main grid is considerably increased when micro grid is connected if the size of the DER is large enough.

5. Impact of appreciable amount of load being met by voltage sourced converters makes the fault currents to be significantly different.

5.2 Over current protection

Over current protection that has been in use for conventional distribution system protection requires some modifications to be made so that it can be used for the protection of mesh connected micro grid with DERs.

In order to understand the proper functioning of the overcurrent protection, let us consider a simple structure of a micro grid shown in **Figure 1**. In general this can be divided into four zones namely MV feeder and busbar protection zone (Z1), transformer protection zone (Z2), LV feeder protection zone (Z3) and micro grid protection zone (Z4) as shown in **Figure 2**.

Based on the location of fault with respect to DER, they can be classified into external (in Z1 and Z2) and internal faults (in Z3 and Z4). If the CB1 is open the micro grid is in islanded mode and if it is closed it is in grid connected mode [18].

5.2.1 Fault outside micro grid

If the fault is in zone Z1 or Z2 then main grid protection system will clear the fault. As per the requirements of IEEE Standard 1547-2003, the micro grid has to be islanded by opening the CB1. If there are inverter based DERs in the micro grid, then the fault current will be limited by them. If conventional over current relays

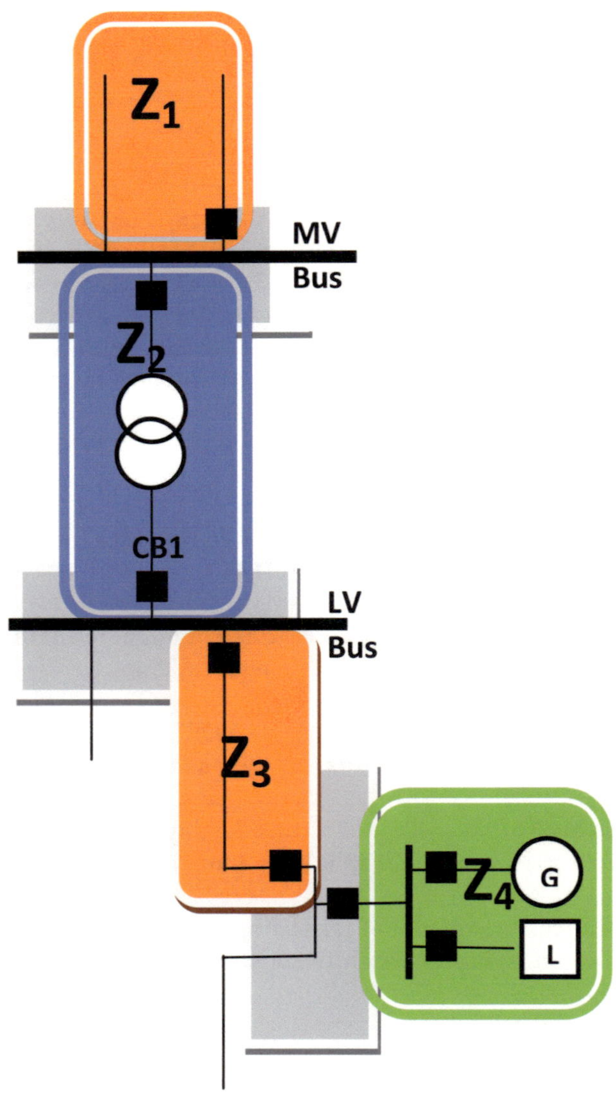

Figure 2.
Typical micro grid showing the zones.

are used for tripping CB1, then the fault current will not be sufficient to trip the breaker. By employing a directional over current relay at LV bus, protection can be ensured. Alternatively changes in frequency or voltage can also be taken as useful indicators for detection of Islanding to initiate the desired protective action. The current setting of this relay should be the cumulative weighted sum of fault current contribution by all the DERs present in the micro grid governed by (Eq. (1)). The weighting factor varies from 1.1 (for inverter based DG) to 5 (for synchronous DG) depending on the number and type of DERs.

$$I_{kmin} = \sum_{1}^{n} k_{DER} * I_{rDER} \tag{1}$$

Here the I_{kmin} is the required adaptive relay current setting, k_{DER} and I_{rDER} are the weighting factor and rated current of the DER [19]. Based on the permissible voltage sag considerations, if sensitive loads are present in micro grid, CB1 should be opened in 70 ms [20].

5.2.2 Fault inside the micro grid

If the fault occurs on the LV feeder or the consumer end, i.e., Z3 or Z4 then the protective system should isolate the faulty section ensuring that minimum number of consumers get affected.

Here again the two cases of grid connected and islanded modes of operation must be considered. Also the presence of inverter based DERs and synchronous based DERs should be given due consideration. Following are the key points to be considered.

- If there is a fault in Z3 or Z4 in grid connected mode, main grid will supply sufficient fault current and faulty section will be isolated.

- If a large synchronous DER is present, then the fault current seen by the relay will be smaller than the fault current without DER causing protection blinding in case of a fault in Z3. It may also lead to delay in tripping the breaker if inverse definite minimum time (IDMT) over current relays are employed for protection. It is due to the fact that the IDMT relay characteristic has inverse characteristic for low magnitude portion of the fault current against the definite time characteristic for higher fault currents.

- A low power diesel generator has low inertia. If there is a delay in the tripping, it might lead to unwanted tripping of the synchronous DER if the power rating is low. To avoid this, a proper adaptive coordination among the relays is essential.

- In islanded mode, if there is fault on Z3 and if there are inverter based DGs, they will limit the fault current as described in the case of faults outside the micro grid earlier.

- In islanded mode, if there is a fault in Z4, it can be isolated by proper relay setting based on the possible fault current supplied by the inverter based DERs without any selectivity problem

In a nutshell, the major challenge in over current protection is the potential difference in the fault currents due to the presence of DERs in grid connected and islanded mode. This calls for adaptive schemes which demand expensive and complex communication infrastructure. The decision of disconnecting/keep it connected/shut down the micro grid depends on several factors such as reliability, cost and the number of customers that get affected. [18] Lot of research is focused on the application of the adaptive over current protection which demands effective communication infrastructure and the IEDs.

5.3 Distance protection

Based on the challenges of relay settings and coordination of the over current relays due the large difference in the fault currents in grid connected and islanded mode, research has been diverted towards the application of distance protection to micro grid in which the tripping decision is based on the impedance seen by the relay and not on the current magnitude [21, 22]. The DER output may result in under reach and power drawn by the loads may cause over reach of the distance relays. By employing more number of distance relays, these issues can be addressed. The impedance seen by the distance relay gets affected by the fault current limiting nature of the inverter based DERs. In case of induction motor generator based DGs employing SCIM (squirrel cage induction motor), when the machine starts absorbing reactive power, the line current leads the voltage. It poses the over reach problem to the connected distance relay which measures it. In case of a DFIM (doubly fed induction motor) based DG, the power factor of the DG unit is controlled by the control system of DFIM during fault conditions. If an unbalanced fault occurs and the fault currents are not large, then the control system can easily maintain the power factor of DFIM. It may lead to protection problems similar to that encountered in case of an inverter based DG [3]. This hinders the application of distance relays for protection of micro grid.

5.4 Differential protection

Difference between the measurements made at different points located in a micro grid (preferably at the two ends of a feeder section) is considered as an actuating quantity for this type of protection. Employing symmetrical components (zero sequence) a differential protection applying the directional features of the difference current can be used in three different ways as shown below [23].

In the first method (shown in **Figure 3**), in order to protect the micro grid and main grid a master micro grid control center (MGCC) is used. Using MGCC it is possible to integrate all protective schemes. Based on the information received from monitoring relays it is expected to protect the main grid and micro grid. However, this method is found to be costly and unreliable to protect either micro grid or main grid alone due to the complex communication infrastructure and the associated data analysis to be carried out.

Second method (shown in **Figure 4**) logic employed only local controllers. Every relay communicates with its neighboring relay directly and monitors the current direction. In this there is no master control center. Whenever a reversal of current is sensed, the faulted section is isolated.

Third method (refer to **Figure 5**) is an improvised version of second method. Each feeder has two monitoring relays. In this method, magnitude of the fault current also is considered in addition to direction unlike the previous two methods. With this the problem of low magnitude fault currents can be handled successfully.

Out of the three methods, second method is more cost effective. In the first method there would be a time delay as the data analysis has to be completed before the protective action is initiated and hence it cannot serve the purpose of primary

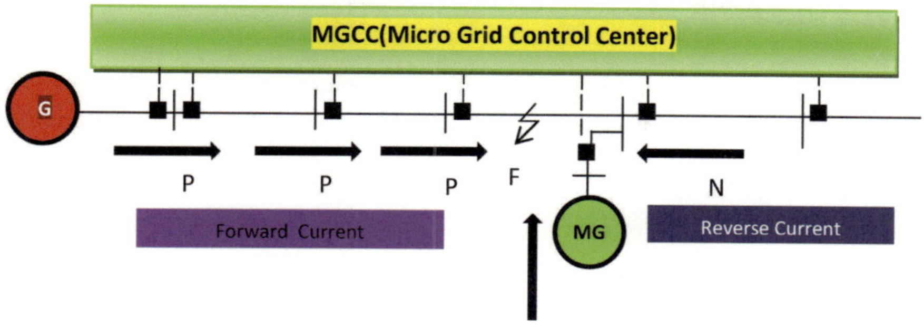

Figure 3.
Method 1 with microgrid control center.

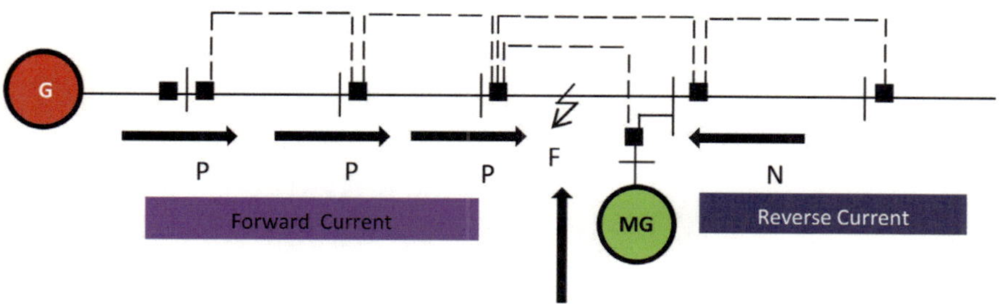

Figure 4.
Method 2 with local controllers.

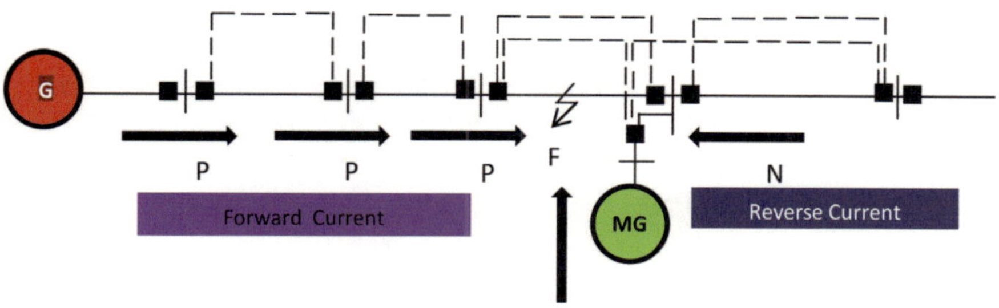

Figure 5.
Method-3 improvized version with local controllers.

protection. In the third method, there is an addition of one more directional monitoring unit in each feeder making it expensive. In all these three methods the fault detection and clearance are reliable and only the faulted section is isolated causing minimum number of consumers to be affected. These schemes do not require any change in the configuration or in the relay settings for both modes of operation of the micro grid and are independent of the type and number of DERs connected to the micro grid [24].

As there are no zero sequence currents in case of a phase to phase fault, negative sequence components of currents are used for fault detection [25]. Using the positive sequence components also considering both amplitude and phase angle the differential protection system is discussed in [26]. However, if there is unbalance and negative and zero sequence currents flow is due to unbalance in the micro grid rather than a fault, these methods need to be examined more carefully. Challenges in this type of protection may be summarized as high cost, communication infrastructure, need for synchronized measurements, effect of unbalanced loads etc.

5.5 Voltage based methodologies

Extensive research has been carried out on these methods initially at University of Bath [27]. In this method voltage is considered for the detection of fault and subsequently for isolation. There are two methods. One is transformation method and the other is harmonic method.

5.5.1 Transformation method

In this method, the output voltage of DER is transformed in two steps. (i) transform voltages from abc to dq frame using Eqs. (2) and (3).

$$\begin{bmatrix} V_{ds} \\ V_{qs} \\ V_0 \end{bmatrix} = \frac{2}{3}\begin{bmatrix} 1 & -1/2 & 1/2 \\ 0 & -\sqrt{3}/2 & \sqrt{3}/2 \\ 1/2 & 1/2 & 1/2 \end{bmatrix}\begin{bmatrix} V_a \\ V_b \\ V_c \end{bmatrix} \tag{2}$$

(ii) From dq transform to dc values

$$\begin{bmatrix} V_{dr} \\ V_{qr} \end{bmatrix} = \begin{bmatrix} \cos\omega t & -\sin\omega t \\ \sin\omega t & \cos\omega t \end{bmatrix}\begin{bmatrix} V_{ds} \\ V_{qs} \end{bmatrix} \tag{3}$$

Any fault condition will get reflected as a change in d-q values.

$$V_{DIST} = V_{qref} - V \tag{4}$$

By comparing with the reference value, it can be easily inferred which type of fault and it can be isolated [27]. Application of transformations is an involved process and becomes complex in certain faults detection. Even a small difference in the voltage drop in case of a short line, shows a considerable effect on protection. Network topology also plays a major role in the application of this method when large numbers of DERs are present.

5.5.2 Harmonic method

In this method, when a fault occurs the total harmonic distortion (THD) of the terminal voltage increases. By comparing the THD of the terminal voltage of the converter with a predefined reference value, the type of fault can be identified. In this method discrete Fourier transforms are employed to convert the phase voltages Va, Vb, Vc into frequency domain. By using proper communication channel between the relays, fault area can be located and isolated [28]. This is used as backup protection. A correct setting for the reference value of THD is often challenging.

5.6 Adaptive protection

In this type of protection, the protection strategy must be modified in line with the existing operating conditions in the micro grid. It is to be done online. To accomplish this, numerical directional O/C relays are a good choice. Existing conventional fuses, electro mechanical and static relays settings and characteristics cannot be changed online. It necessitates that the existing protection equipment be upgraded to meet the requirement. Complying to IEC 61850 and installation of IEDs (Intelligent Electronic Devices) at appropriate places can make the relays to be adaptive with the ability to adjust their settings and characteristics accordingly on receiving the signals online or following a time sequence. Thorough study of all possible topological configurations is to be carried out offline prior to the operation. It also necessitates conducting power flow studies and carrying out short circuit analysis for each configuration that might occur. For adjusting the settings and characteristics, fast and effective communication infrastructure should be in place [29].

5.7 Methods of improving protection

Considerable changes in fault current magnitudes during the grid connected and islanded modes of operation calls for alternative measures to be taken to improve the protection. If it is possible to modify the fault current magnitude whenever there is a change of operating mode of the micro grid, the existing protective systems can be used with some changes without the need of replacing them. If the fault current can be modified suitably by deploying some additional components, it would be very useful. These may be used either to increase or decrease the fault current suitably to have correct protective action along with the coordination among different protective equipment used. Response of a synchronous DER is different from an inverter fed DER during fault conditions. In case of an inverter fed DER, fault current need to be increased and in case of synchronous DG it should be reduced. Usage of fault current limiters (FCL), employing an interfacing unit at the point of micro grid interconnection with main grid to avoid the fault feeding from main grid are some of the available options. These options demand huge investment and maintenance. They depend on the proper functioning of islanding detection methods employed. Fault current limiting poses challenges if the size and penetration level of the DERs is high.

5.8 Protocols and standards

Some of the relevant standards related to micro grid operation are listed here for reference. IEEE Standard 1547 series covers Standard for Interconnecting Distributed Resources with Electric Power Systems. Standard Conformance Test Procedures for Equipment Interconnecting Distributed Resources with Electric Power Systems are given by IEEE Std 1547.1, Guide for Monitoring, Information Exchange, and Control of Distributed Resources Interconnected with Electric Power Systems is presented in IEEE Std 1547.3. Guide for Design, Operation, and Integration of Distributed Resource Island Systems with Electric Power Systems is IEEE Std 1547.4 and recommended Practice for Interconnecting Distributed Resources with Electric Power Systems Distribution Secondary Network are presented in IEEE Std 1547.6 [30]. There are reports prepared by CIGRE Working group also for reference. WG C6.22: Micro grid Evolution Roadmap contains the definitions and nomenclature of micro grid, WG C6.24 explains Capacity of Distribution Feeders for Hosting DER Connection and Integration of DER [31].

5.9 Recommendations

Of all the protection methods discussed above, differential current relaying is the most suited protection system for micro grid. This will enable fault location and also clearing in minimal time. Either one can have pilot lines for connecting the relays differentially (normally the feeder length in Distribution systems is not high so the cost of pilot lines also will be low) or one can locate RTUs at the two ends of each feeder and they will communicate the current magnitude, phase and direction to a central station where the fault location and tripping decisions are taken. The current measured at each end of the feeder is applied to Directional Over current relay with fixed operating time to provide backup protection. This system will not require any changes either in configuration or settings for faults in the micro grid or in the main grid. Also this is not affected by the number and location of DERs and whether the micro grid is connected or isolated from the main grid.

5.10 Detection of islanding

It plays a major role in proper functioning of protective system. If differential relaying is adapted there is no necessity to detect islanding. However for the Control of micro grid operation and to maintain the power quality the control system for each DER has to be changed since the reference signal for frequency and voltage which is taken from the grid will not be available when the micro grid is isolated. There are different methods available in the literature for islanding detection such as rate of change of frequency, voltage, power factor, THD. Also the use of FFT or Wavelet transform of the terminal voltage will give out different spectrum when isolation takes place. Artificial intelligence techniques also have been employed for detection of islanding. Some new hybrid techniques employing these techniques can be found in Refs. [32–35].

Year	Title	Ref No.	Methodology	Type of faults discussed	Micro grid features	Remarks
2004	Brahma SM, Girgis AA. Development of adaptive protection scheme for distribution systems with high penetration of distributed generation. IEEE Transactions on Power Delivery. 2004;**19**(1):56-63	[36]	Here protection scheme is developed for micro grids with Synchronous DERs operating in grid connected mode addressing the fuse to fuse, fuse to recloser co-ordination issues that arises due to large number of DERs. The relaying strategy is adaptable in view of temporary faults and permanent faults and extension of the scheme to additional feeders.	Balanced and unbalanced faults	(i) Grid connected mode (ii) Radial system (iii) Synchronous based DER	(i) Applicable only in grid connected mode (ii) Protection in the islanded mode of operation is not included (iii) Works well when large number of DERs are connected in the micro grid. If the number DERs is less, it poses challenges.
2005	Wan H, Li KK, Wong KP. A multi-agent approach to protection relay coordination with distributed generators in industrial power distribution system. In: Fortieth IAS Annual Meeting. Conference Record of the 2005 Industry Applications Conference, 2005. Vol. 2. IEEE; 2005. pp. 830-836	[37]	Protective relay coordination using a multi agent communication approach is presented. It is capable of providing back up protection in case of primary protection failure in grid connected mode is developed. Make use of the Java agent Development Framework (JADE) platform for simulation of communication.	Fault type is not specified	(i) Grid connected mode (ii) Radial system (iii) Both synchronous based and inverter based DERs	(i) Applicable only in grid connected mode (ii) Relay coordination is dependent on communication (iii) Capable of providing backup protection

Year	Title	Ref No.	Methodology	Type of faults discussed	Micro grid features	Remarks
2006	Al-Nasseri H, Redfern MA, Li F. A voltage based protection for micro-grids containing power electronic converters. In: IEEE Power Engineering Society General Meeting; 2006. p. 7	[27]	Here DER output voltage transformation from abc to dq frame is performed and then the deviations of these values from reference values are computed. Based on the difference the protective action is initiated. A communication link is provided between relays. (Voltage based protection schemes)	Balanced and unbalanced faults	(i) Islanded mode (ii) Radial system (iii) Inverter based DER (iv) Constant MVA load (v) Overhead line with voltage level 11 kV/0.48 kV	(i) Protection against high impedance faults is not considered (ii) Effect of single pole tripping is not explained (iii) Relay functioning depends on the communication link between the relays
2006	Perera N, Rajapakse AD, Agent-based protection scheme for distribution net-works with distributed generators. In: IEEE Power Engineering Society General Meeting; 2006. p. 6	[38]	Network is divided into several segments. Relay agents communicate through an asynchronous communication link. Time domain simulation is done using wavelets for fault location. Central data processing is not required as the decisions are done in a distributed manner.	Phase to ground and phase to phase to ground fault high impedance fault	(i) Both grid-connected and islanded mode (ii) D-DGs (iii) Constant MVA Load (iv) OHL radial 24.9 kV	(i) Requires only current measurements and these measurements need not be time synchronized (ii) Demands high speed communication for proper determination of fault section (iii) Poses challenges to avoid relay functioning during switching transients
2007	Nikkhajoei H, Lasseter RH. Microgrid protection. In: IEEE Power Engineering Society General Meeting; 2007. pp. 1-6	[17]	A static switch is placed at the point of common coupling. Entire system is divided into different zones. Makes use of symmetrical components and system residual current is used for protective action.	Phase to ground and phase to phase faults	(i) Islanded mode (ii) Radial system (iii) Inverter based DER (iv) kW load (v) 0.48 kV distribution voltage	(i) Protection against high impedance faults is not considered (ii) Effect of single pole tripping is not explained (iii) Three phase faults are not discussed
2008	Al-Nasseri H, Redfern MA. Harmonics content based protection scheme for micro-grids dominated by solid state converters. In: 12th International Middle-East Power System Conference, 2008 (MEPCON 2008); 2008. pp. 50-56	[28]	Protection System is based on the measurement of amount of harmonic content present during the fault condition. For each type of fault a threshold value of THD is evaluated and set as a reference. Based on the measured value of harmonic content, required protective action will be initiated. (Voltage based protection schemes)	Balanced and unbalanced faults	(i) Islanded mode (ii) Radial system (iii) Inverter based DER (iv) Constant MVA load (v) Overhead line with voltage level 11 kV/0.48 kV	(i) It is required to assess the reference THD values for different fault scenarios which would be challenging (ii) If any DER supplies a harmonic free voltage or with lesser harmonic content, protection system may fail. (iii) Variable fault impedances, large dynamic load switching poses sensitivity issues demand for proper settings of threshold limits of THD

Year	Title	Ref No.	Methodology	Type of faults discussed	Micro grid features	Remarks
2009	Dewadasa M, Ghosh A, Ledwich G. An inverse time admittance relay for fault detection in distribution networks containing DGs. In: 2009 IEEE Region 10 Conference (TENCON 2009); 2009. pp. 1-6	[39]	These relays have the ability to operate for faults in both forward direction and reverse direction. Its operation is based on the measured admittance and has an inverse time characteristic. The protection system can operate for low fault currents also and thus provide protection under islanded mode also. It is possible to supply the load in islanded mode also. Network is divided into different zones .	Balanced and unbalanced faults	(i) Both grid-connected and islanded mode (ii) Inverter based-DGs (iii) Constant MVA (iv) OHL radial and closed loop (v) 11 kV	(i) fundamental frequency component extraction may lead to measurement errors due to harmonics and dc offset (ii) Takes more time of operation for high impedance faults (iii) Does not use any communication link
2010	Sortomme E, Venkata M, Mitra J. Microgrid protection using communication-assisted digital relays. In: IEEE PES General Meeting; Providence, RI; 2010. p. 1	[40]	In this method, digital relays are employed along with communication network. An additional line is added in the system to simulate the loop structure in this paper. A new modeling for high impedance fault simulation is presented.	Balanced and unbalanced faults	(i) Grid connected and islanded mode (ii) Both inverter and synchronous based DGs (iii) Radial and loop structure (iv) 18 bus system with multiple DGs included (v) Unbalanced load is also included	(i) Highly expensive and time synchronization is not considered (ii) Imbalance created between generation and demand due to line removal in the radial mode makes the protection challenging and calls for effective communication infrastructure and sensors. (iii) In case of communication failure, protection against high impedance faults is at stake
2010	Shi S, Jiang B, Dong X, Bo Z. Protection of microgrid. In: 10th IET International Conference on Developments in Power System Protection (DPSP 2010); Managing the Change; 2010. pp. 1-4	[41]	This protection scheme is based on current travelling waves. Here detection of the faults is done using busbar voltages and location of the fault is found out employing current travelling waves. No communication link is used. Based on the information available locally, protective relay works.		Both grid-connected and islanded mode - - - 10/0.4 kV distribution voltage	(i) Method is independent of unbalance between the load and generation, level of fault current or power flow (ii) Simulation results are not presented

Year	Title	Ref No.	Methodology	Type of faults discussed	Micro grid features	Remarks
2011	Voima S, Kauhaniemi K, Laaksonen H. Novel protection approach for MV microgrid. In: CIRED 21st International Conference on Electricity Distribution; 6–9 June, 2011; Frankfurt; 2011. Paper No. 0430	[42]	This is an adaptive protection scheme which uses tele-communication infrastructure. Network is divided into four different zones. IEDs used have directional over current protection function along with current and voltage measurements. To achieve proper selectivity, interlocking signal is sent along with the direction of fault. Applicability of distance relay also is presented.	Specific type of fault details are not mentioned	(i) Islanded mode (ii) Radial system (iii) Inverter based DER (iv) Constant MVA load (v) Over head line with voltage level 20 kV	(i) High dependency on the communication infrastructure (ii) With reliable communication links it can be made adaptable to different modes of operation (iii) Details of simulation of different fault scenarios is missing
2012	Samantaray SR, Joos G, Kamwa I. Differential energy based microgrid protection against fault conditions. In: IEEE PES Innovative Smart Grid Technologies (ISGT); 2012. pp. 1-7	[43]	In this method, differential energy applying time frequency transform is used to initiate the protective action. On either end of the feeder, amount of spectral energy is found out. High impedance faults are also considered.	Balanced and unbalanced faults	(i) Both grid-connected and islanded mode (ii) Both inverter and grid connected mode (iii) Constant MVA (iv) OHL radial and closed loop 25 kV distribution voltage	(i) Differential energy is used to recognize the fault patters (ii) Makes use of both time and frequency data where as in other schemes only one data is used. (iii) Setting the threshold limit for the differential energy plays crucial role
2013	Ustun TS, Ozansoy C, Ustun A. Fault current coefficient and time delay assignment for microgrid protection system with central protection unit. IEEE Transactions on Power Systems. 2013;**28**:598-606	[44]	In this method communication based coordination has been presented. Amount of fault current contribution by any DG is represented as a coefficient. Selectivity of the relays is controlled by automatic adjustment of the current setting.	Balanced faults	(i) Grid connected and islanded mode (ii) Inverter based and synchronous based DGs (iii) Radial system	(i) Requires human input however it can be minimized if the structure of the network is obtained by running an automated algorithm (ii) Delay in the communication depends on the type of protocol used (iii) faults within the micro grid only are considered
2014	Kar S, Samantaray SR. Time-frequency transform-based differential scheme for microgrid protection. IET Generation, Transmission & Distribution. 2014;**8**:310-320	[45]	The protection scheme identifies the fault current patterns based on the S transforms. Differential energy is computed considering both ends of the feeder and it is used for protective action.	Balanced and unbalanced faults	(i) Both grid-connected and islanded mode (ii) Inverter based and synchronous DGs (iii) Constant MVA load (iv) OHL radial and closed loop 25 kV	(i) Results are compared with the current differential technique for all fault scenarios (ii) Differential energy is less sensitive to time synchronization errors compared to current difference

Year	Title	Ref No.	Methodology	Type of faults discussed	Micro grid features	Remarks
2015	Kanakasabapathy P, Mohan M. Digital protection scheme for microgrids using wavelet transform. In: 2015 IEEE International Conference on Electron Devices and Solid-State Circuits (EDSSC). IEEE; 2015. pp. 664-667	[46]	Using wavelet transforms a microprocessor based protection scheme is developed for grid connected mode of the micro grid with fault detection and classification. Cumulative sum of the high frequency details of power signal is computed and is compared against a threshold value to send the trip signal using the digital relay.	Balanced and unbalanced faults	(i) Grid connected and disconnected mode (ii) Radial system (iii) Both Synchronous based and inverter based DERs	(i) Fault location depends on the power signal high frequency details (Phfd) (ii) Threshold value of Phfd depends on sampling frequency of the analog signal and the type of wavelet chosen
2016	Gururani A, Mohanty SR, Mohanta JC. Microgrid protection using Hilbert–Huang transform based-differential scheme. IET Generation, Transmission & Distribution. 2016;**10**(15):3707-3716	[47]	Hilbert-Huang transform (HHT) has been employed to determine the differential energy in this method. To discriminate faults in islanded mode and in case of high impedance fault an appropriate setting for the differential energy is used as threshold value.	Balanced and unbalanced faults	(i) Both grid connected and islanded modes (ii) Radial system (iii) Inverter based DERs	(i) Setting of proper threshold value is important to discriminate different fault conditions (ii) When noise is included in the signals protection becomes challenging
2017	Hooshyar A, Iravani R. Microgrid protection. Proceedings of the IEEE. 2017;**105**(7):1332-1353	[3]	A comprehensive review of the micro grid protection techniques has been presented along with several case studies using different relays in different modes of operation employing synchronous based and inverter based DGs. The fault ride through capability also is discussed. DC microgrid protection is also discussed briefly.	Balanced and unbalanced faults	(i) Both grid connected and islanded modes (ii) Radial system and mesh system, voltage 12.47 kV (iii) Inverter based and synchronous based DERs	(i) DG ride through capabilities in islanded mode for different fault scenarios is presented. (ii) Effect of ECDG units on directional over current relays and distance relays is shown to be more than in case of differential relays. (iii) Frequency of fault current is shown to be dependent on the slip of induction machine. (iv) In case of DFIG based microgrid, response of the relay is shown to be dependent on the type of control strategy employed

Year	Title	Ref No.	Methodology	Type of faults discussed	Micro grid features	Remarks
2018	Aghdam TS, Karegar HK, Zeineldin HH. Variable tripping time differential protection for microgrids considering DG stability. IEEE Transactions on Smart Grid	[48]	This method discusses the stability aspect also along with fault clearance. A multi agent approach along with the zoning principle is employed. For coordination and backup purposes each agent has three layers namely primary, backup and bus protection. The critical clearing time (CCT) curves of the DGs employed are analysed to establish the mechanism for checking the constraints on the CCT are developed.	Balanced fault	(i) Both grid connected and islanded mode (ii) Modified CIGRE benchmark micro grid test system (iii) Synchronous DG (iv) 20 kV.	(i) Micro grid has only synchronous DGs (ii) Specific type of fault is not mentioned (iii) Fuse tripping slower for the same fault current as the nominal current of the fuse increase. (iv) Settings of several differential layers depend on the fuse size for coordination.

Table 1.
Summary of research on protection of micro grids.

Table 1 gives a consolidated picture of ongoing efforts for protection of Micro Grids.

6. Conclusions

A comprehensive review of various protection methods as applicable to micro grid protection is presented. DERs are becoming an integral part of distribution systems but the adequate changes necessary in the protection system has not yet picked up the pace. Lot of research is going on in this area to use the existing protective infrastructure justifiably without compromising on the safety aspect. It is apparent that well-built communication infrastructure is essential for meeting the requirement of micro grid protection. It is due to the fact that there are inevitable topological changes in the network due to the transition of micro grid operating mode from grid connected to islanded and vice versa. Also, the intermittent nature of the DER output and the fault limiting features of inverter fed DGs present several technical challenges to the micro grid protection engineers. Making the protective system to be adaptive is the need of the hour. But it involves lot of infrastructure development and is costly. Many methods based on directional O/C relays, distance relays and voltage based protection schemes have been proposed for effective implementation. However, effective utilization of the existing protective systems with minimal changes in the infrastructure appears to be possible with differential protection scheme. With the advancements in communication technology, micro grid protection can be made adaptive in a cost effective manner.

Author details

Mylavarapu Ramamoorty and Suraparaju Venkata Naga Lakshmi Lalitha*
KLEF Deemed to be University, Guntur, Andhra Pradesh, India

*Address all correspondence to: lalitha@kluniversity.in

References

[1] Marnay C et al. Microgrid evolution roadmap. In: 2015 International Symposium on Smart Electric Distribution Systems and Technologies (EDST); Vienna; 2015. pp. 139-144

[2] Benchmark Systems for Network Integration of Renewable and Distributed Energy Resources. Taskforce C6.04.02. CIGRE-2014

[3] Hooshyar A, Iravani R. Microgrid protection. Proceedings of the IEEE. 2017;**105**(7):1332-1353

[4] Chowdury S, Chowdury SP, Crossley P. Microgrids and Active Distribution Networks. Vol. 4-5. London, United Kingdom: The Institution of Engineering and Technology; 2009. pp. 70-95

[5] Memon AA, Kauhaniemi K. A critical review of AC Microgrid protection issues and available solutions. Electric Power Systems Research. 2015;**129**:23-31

[6] Shiles J et al. Microgrid protection: An overview of protection strategies in North American microgrid projects. 2017 IEEE Power & Energy Society General Meeting; Chicago, IL; 2017. pp. 1-5

[7] Mason CR. The Art and Science of Protective Relaying. New York: John Wiley & Sons; 1956

[8] Van C. Warrington AR. Protective Relays, Their Theory and Practice. Vol. II. London: Chapman & Hall; 1974

[9] Sharaf HM, Zeineldin HH, El-Saadany E. Protection coordination for microgrids with grid-connected and islanded capabilities using communication assisted dual setting directional overcurrent relays. IEEE Transactions on Smart Grid. 2018;**9**(1):143-151

[10] Wheeler KA, Elsamahy M, Faried SO. A novel reclosing scheme for mitigation of distributed generation effects on overcurrent protection. IEEE Transactions on Power Delivery. 2018;**33**(2):981-991

[11] Ustun TS. Interoperability and interchangeability for microgrid protection systems using IEC 61850 standard. In: 2016 IEEE International Conference on Power and Energy (PECon); Melaka; 2016. pp. 7-12

[12] Ustun TS, Ozansoy C, Zayegh A. Modeling of a centralized microgrid protection system and distributed energy resources according to IEC 61850-7-420. IEEE Transactions on Power Systems. 2012;**27**(3):1560-1567

[13] Kang X, Nuworklo CEK, Tekpeti BS, Kheshti M. Protection of micro-grid systems: A comprehensive survey. The Journal of Engineering. 2017;**2017**(13):1515-1518

[14] Adamiak M, Baigent D, Mackiewicz R. IEC 61850 Communication Networks and Systems in Substations - An Overview for Users. GE Grid Solutions Protection & Control Journal. 8th Edition, Spring; 2009. pp. 61-68. https://www.gegridsolutions.com/multilin/journals/issues/spring09/iec61850.pdf

[15] IEEE Standard for Interconnecting Distributed Resources with Electric Power Systems. In: IEEE Std 1547-2003. 28 July 2003:1-28

[16] Conti S, Raiti S. Integrated protection scheme to coordinate MV distribution network devices, DG interface protections and microgrid operation. In: 2009 International Conference on IEEE Clean Electrical Power; 2009. pp. 640-646

[17] Nikkhajoei H, Lasseter RH. Microgrid protection. In: IEEE Power Engineering Society General Meeting; 2007. pp. 1-6

[18] Advanced Architectures and Control Concepts for MORE MICROGRIDS: STREP project funded by the EC under 6FP, Contract No: SES6- 019864. Available from: http:// www. microgrids.eu/documents/668. pdf

[19] The Contribution to Distribution Network Fault Levels From the Connection of Distributed Generation. DG/CG/00027/00/00 URN. by KEMA Limited. 2005. Available from: https:// webarchive.nationalarchives. gov. uk/20100919182422/; http:// www. ensg.gov.uk/assets/14_06_2005_ dgcg0000200.pdf

[20] EPRI. Overview of SEMI F47-0706. 2006. Available from: www.f47testing. com

[21] IEEE Guide for Design, Operation, and Integration of Distributed Resource Island Systems with Electric Power Systems. In: IEEE Std 1547.4-2011. 20 July 2011:1-54

[22] Sinclair A, Finney D, Martin D, Sharma P. Distance protection in distribution systems: How it assists with integrating distributed resources. IEEE Transactions on Industry Applications. 2014;**50**(3):2186-2196

[23] Yan-xia Z, Feng-xian D. New schemes of feeder protection for distribution networks including distributed generation. Automation of Electric Power Systems. 2009;**33**(12):71-74

[24] Jiang W, He Z, Bo Z. The overview of research on microgrid protection development. In: 2010 International Conference on Intelligent System Design and Engineering Application; Changsha; 2010. pp. 692-697

[25] Nikkhajoei H, Lasseter RH. Microgrid Fault Protection Based on Symmetrical and Differential Current Components. Wisconsin Power Electronics Research Center, Department of Electrical and Computer Engineering. University of Wisconsin-Madison; 2006

[26] Ling-Ling Z, Chang-Kai L, Hua-Zhong Z, Pei-yi Z. Directional overcurrent protection for distribution systems containing distributed generation. Power System Technology. 2009;**33**(14):94-98

[27] Al-Nasseri H, Redfern MA, Li F. A voltage based protection for micro-grids containing power electronic converters. In: IEEE Power Engineering Society General Meeting on 2006; 2006. pp. 1-7

[28] Al-Nasseri H, Redfern MA. Harmonics content based protection scheme for Micro-grids dominated by solid state converters. 2008 12th International Middle-East Power System Conference, Aswan; 2008. pp. 50-56

[29] Habib HF, Lashway CR, Mohammed OA. A review of communication failure impacts on adaptive microgrid protection schemes and the use of energy storage as a contingency. IEEE Transactions on Industry Applications. 2018;**54**(2):1194-1207

[30] Thomas B. IEEE 1547 and 2030 Standards for Distributed Energy Resources Interconnection and Interoperability with the Electricity Grid. Technical report NREL/TP-5D00-63157 December 2014. Contract No. DE-AC36-08GO28308. Available from: https://www.nrel.gov/docs/ fy15osti/63157.pdf

[31] Network of the Future. Electricity Supply Systems of the Future. Final report by CIGRE WG. April 2011. Available from: https://ec.europa.eu/assets/jrc/events/20130926-eco-industries/20130926-eco-industries-hatziargyriou.pdf

[32] Mohanty SR, Kishor N, Ray PK, Catalo JPS. Comparative study of advanced signal processing techniques for islanding detection in a hybrid distributed generation system. IEEE Transactions on Sustainable Energy. 2015;**6**(1):122-131

[33] Pigazo A, Liserre M, Mastromauro RA, Moreno VM, Dell'Aquila A. Wavelet-based islanding detection in grid-connected PV systems. IEEE Transactions on Industrial Electronics. 2009;**56**(11):4445-4455

[34] Khodaparastan M, Vahedi H, Khazaeli F, Oraee H. A novel hybrid islanding detection method for inverter-based DGs using SFS and ROCOF. IEEE Transactions on Power Delivery. 2017;**32**(5):2162-2170

[35] Kermany SD, Joorabian M, Deilami S, Masoum MAS. Hybrid islanding detection in microgrid with multiple connection points to smart grids using fuzzy-neural network. IEEE Transactions on Power Apparatus and Systems. 2017;**32**(4):2640-2651

[36] Brahma SM, Girgis AA. Development of adaptive protection scheme for distribution systems with high penetration of distributed generation. IEEE Transactions on Power Delivery. 2004;**19**(1):56-63

[37] Wan H, Li KK, Wong KP. A multi-agent approach to protection relay coordination with distributed generators in industrial power distribution system. In: Conference Record of the 2005 Industry Applications Conference2005. Fortieth IAS Annual Meeting. Vol. 2. IEEE; 2005. pp. 830-836

[38] Perera N, Rajapakse AD. Agent-based protection scheme for distribution net-works with distributed generators. In: IEEE Power Engineering Society General Meeting; 2006. p. 6

[39] Dewadasa M, Ghosh A, Ledwich G. An inverse time admittance relay for fault detection in distribution networks containing DGs. In: IEEE Region 10 Conference (TENCON 2009); 2009. pp. 1-6

[40] Sortomme E, Venkata M, Mitra J. Microgrid protection using communication-assisted digital relays. In: IEEE PES General Meeting; Providence, RI; 2010. p. 1

[41] Shi S, Jiang B, Dong X, Bo Z. Protection of microgrid. In: 10th IET International Conference on Developments in Power System Protection (DPSP 2010); Managing the Change; 2010. pp. 1-4

[42] Voima S, Kauhaniemi K, Laaksonen H. Novel protection approach for MV Microgrid. In: CIRED 21st International Conference on Electricity Distribution; 6-9 June, 2011; Frankfurt; 2011. Paper No. 0430

[43] Samantaray SR, Joos G, Kamwa I. Differential energy based microgrid protection against fault conditions. In: IEEE PES Innovative Smart Grid Technologies (ISGT); 2012. pp. 1-7

[44] Ustun TS, Ozansoy C, Ustun A. Fault current coefficient and time delay assignment for microgrid protection system with central protection unit. IEEE Transactions on Power Systems. 2013;**28**:598-606

[45] Kar S, Samantaray SR. Time-frequency transform-based

differential scheme for Microgrid protection. IET Generation Transmission and Distribution. 2014;**8**:310-320

[46] Kanakasabapathy P, Mohan M. Digital protection scheme for microgrids using wavelet transform. In: 2015 IEEE International Conference on Electron Devices and Solid-State Circuits (EDSSC); IEEE; 2015. pp. 664-667

[47] Gururani A, Mohanty SR, Mohanta JC. Microgrid protection using Hilbert–Huang transform based-differential scheme. IET Generation Transmission and Distribution. 2016;**10**(15):3707-3716

[48] Aghdam TS, Karegar HK, Zeineldin HH. Variable tripping time differential protection for microgrids considering DG stability. IEEE Transactions on Smart Grid. May 2019;**10**(3):2407-2415

Chapter 6

Innovative Differential Protection Scheme for Microgrids Based on RC Current Sensor

Ali Hadi Abdulwahid and Adnan A. Ateeq

Abstract

The modern power system and future ones include several intelligent devices. It also integrates renewable energy sources, energy storage, energy microgrid control system, hybrid networks, and smart grids with the wide application of information technology and communication. The most crucial goal in the smart grid application is improving safety reliability within the network. Recent studies have recommended the development of smart grid technology that enhances the reliability of electric power systems increases efficiency and improves the detection of faults for protection; this will reduce the duration of interruption of the number of customers affected by the outages. Moreover, smart grid technology decreases the power loss of energy usage and improves the efficiency of the system. And protection is one of the most important challenges facing smart grid deployment. In this chapter, protection for smart grids using differential relays is presented. The differential scheme is a very reliable method of ensuring the safety of protected areas. This chapter discusses the differential relay parameters with various fault conditions. Therefore, the protection scheme affirms the rapid separation of the fault zone to reduce damage to the equipment. The simulation results show that the method is effective and reliable.

Keywords: microgrid, reliability, renewable energy sources, power grids, power distribution, differential protection

1. Introduction

The latest revolution in the electricity network technology is called smart grid (SG). Smart grid term describes the future intelligent electric power that using digital technology to monitor and control. In other words, smart grids integrate information and communication intelligently to improve the electricity delivered to customers. It will also enhance safety and reliability, and financial control services in the system. A smart grid is a version of the future power grid that employs advanced equipment and services together with intelligent monitoring, control, communication, and intellectual protection. It is referred to as a revolution in the future of electric power grids because by using applicable technologies, it is a modern and integrated system. With increased energy demands and the expansion of renewable energy sources, power grid systems must be moderated and improved. The smart grid will integrate all types of electric power sources, and accommodate all means of energy generation and distribution to meet the future demands of energy and its technologies [1].

Quality of power delivery is a significant goal of the smart grid that will provide a variety of needs and options at different costs. Furthermore, smart grids will provide advanced monitoring and control by employing intelligent equipment such as digital sensors, electronic switches, smart energy metering, and creative and advanced communication systems. Its data acquisition and control systems include interactive software, real-time control, and power flow analysis. All different types of renewable energy sources will be interconnected with the energy grid system to improve quality, reliability, and stability by using intelligent and advanced devices. Providing advanced technology such as the smart grid requires a smart and intelligent protection system to improve the efficiency of power delivery to customers, and to reduce outages. Employing the smart grid allows energy consumers to be active participants by providing information and options to control the electric demand balance [2].

The microgrid is used to provide customers with economical and reliable power resources and to make effective use of them through the formation of a smart grid structure during the disturbance. However, the protection of microgrid is a challenging task [3–5]. This chapter discusses the application of differential protection schemes. Issues related to protection include bidirectional power flow; it also handles the decrease in fault current levels [6–8]. The power system must operate safely at all times. The main requirements for power system protection include speed, selectivity, sensitivity, safety, reliability and dependability. The reliability requirements of the protection system ensure that appropriate and operable protective measures taken even when certain parts of the protective device may fail [9].

2. Challenges in implementation of smart grids

As shown in **Figure 1**, the key features of the smart grid offer many advantages and prospects in the power industry, thereby revitalizing the socioeconomic strategies of this sector. However, the extensive applications of emerging technologies, if not considered, have vulnerabilities that may result in disasters such as long-term blackouts, economic failure, and so on. **Table 1** provides a brief survey of some of the challenges of smart grid technology [10].

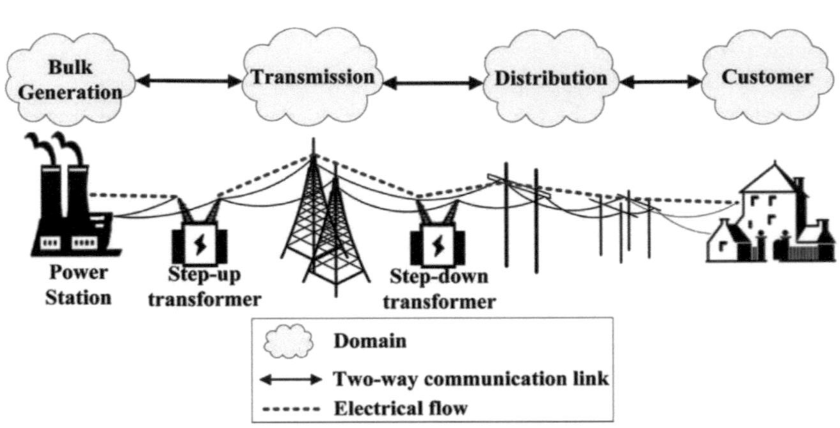

Figure 1.
Future smart grid processing technology.

Technology	Challenges	Issues
Self-healing action	Security	Exposed to Internet attacks (spams, worms, virus, and others), question of national security
	Reliability	Failure during natural calamities, system outages, and total blackout
Renewable energy integration	Wind/PV generation and forecasting	Long-term and unpredictable intermittent sources of energy, unscheduled power flow and dispatch
	Power flow optimization	Transmission line congestion and huge investments
	Power system stability	Decoupling causes system stability issues and causes reduced inertia due to high level of wind penetration
Energy systems storage	Complexity	Complex customary design module and networks
	Non-flexibility	Unique designs for all individual networks; no ease of adaptation
Consumer motivation	Security	Malware, data interception, data corruption, illegal power handling, and smuggling
	Privacy	Sharing of data causes privacy invasion, identity spoofing, eavesdropping, and other problems
Reliability	Grid automation	Need for strong data-routing system with secure and private network for reliable protection, control, and communication
	Grid Reconfiguration	Generation demand equilibrium and power system stability with grid complexity
Power quality	Disturbance identification	System instability during sags, dips, or voltage variation such as over-voltages, under-voltages, voltage flickers, and other problems

Table 1.
Challenges of smart grid technology.

Extensive research on this technology, which aims to overcome many challenges, has been launched by various universities. The system parameters and configuration of the power grids were investigated more intelligently.

3. Rogowski coil current sensor

The Rogowski coil (RC) current sensor works in the same way as a conventional AC core current transformer. They are not closed loops, making the coils open and flexible, and can be wrapped around the conductors [11–13]. RCs can immediately respond to the changing currents, down to a few nanoseconds, due to their low inductance. RC has no iron core to saturate, making it highly linear even when exposed to large currents. Linearity also allows high-current RCs to be determined using smaller reference currents. No danger is observed in opening the secondary winding [14]. Power construction costs and temperature compensation are simple [15–17]. Besides, RC does not use the magnetic core to support two windings. RC is designed with two coils that are electrically connected in the opposite direction, eliminating the electromagnetic field from the outer ring routing. For obtain the current sensor quality, the RC design must meet two important criteria; the first criterion, the mutual inductance M must have a constant value for any of the main conductor locations inside the coil loop, which can be achieved if the coil core has a constant cross section S, is perpendicular to the median line and is constructed with a constant turn density n. Second, the effects of adjacent conductors carrying a large current to the coil of the output signal should be minimal. The following formula defines mutual inductance M:

$$M = \mu_o \cdot n \cdot S \tag{1}$$

When the core has a constant cross-section S, μ_o is the permeability of the free space, and the winding line is perpendicular to the midline m, with a constant density n. The output voltage is proportional to the measured rate of change of current, as shown in Eq. (2) [18]:

$$v(t) = -\mu_o nS\frac{di(t)}{dt} = -M\frac{di(t)}{dt} \tag{2}$$

Where M is the mutual inductance of the coil, also called the sensitivity of the RC. The CT iron core has a nonlinear characteristic and is therefore saturated when a high current or a direct current component is present in primary current. When CT is saturated (that is, the CT ratio error increases), which adversely affects the performance of the relay. **Figure 2** displays the equivalent circuit of a current transformer.The current phase angle between the primary coil and the secondary voltage is almost 90° (due to the coil inductance Ls). **Figure 3** shows the equivalent circuit of RCs.

Rogowski coils are linear and can be used in measurement applications. The oscillatory response of *RC* can be represented by voltage response and natural frequency, as described below:

$$V_o(t) = -M\frac{di(t)}{dt} = -M\frac{di(t)}{dt}.e^{-\xi\omega_n t}\sin\left(\omega_n\sqrt{1-\xi^2}\right)t, \tag{3}$$

where ω_n is the natural frequency, and ε is the damping coefficient. As a result, RCs can replace conventional CTs for measurement and protection. IEEE Std C37.235TM-2007 [19] provides guidance on the application of RC sensors when used for protection purposes, review the essential characteristics.

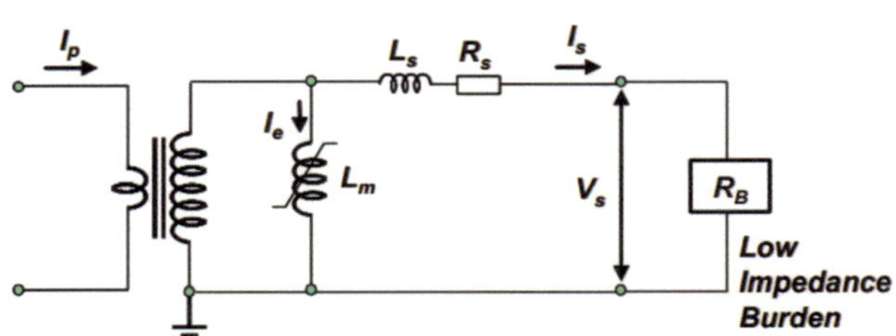

Figure 2.
Current transformer equivalent circuit.

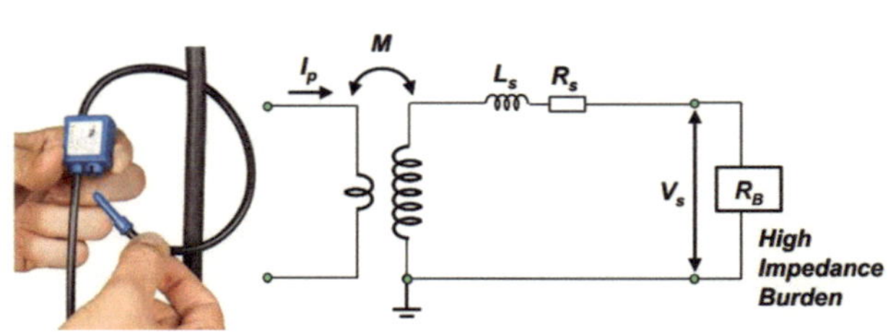

Figure 3.
Rogowski coil equivalent circuits.

- Accuracy: Rogowski coils can be used for measurement, protection, and control. Traditionally, separate secondary windings are used for measurement and protection. The linearity range of sensors enables the combination of sensors for metering and protection in a single device, thereby resulting in sensors with smaller dimensions.

- Linearity (no saturation): Rogowski coils are linear over a wide range of currents. No calculation specific to the control for different primary currents is required. Calculation of precision limit factors is also unnecessary.

- No accuracy versus burden calculation: Rogowski coils are used with microprocessors that have high input impedance. Therefore, calculating the accuracy and burden (no precision and load calculation) is not needed.

- Size: Rogowski coil devices are compact and can be easily combined with voltage sensors in one device, which is known as Combi sensor. Furthermore, RCs can be integrated into other equipment such as circuit breakers, power transformers, and switches.

- Weight: Rogowski coils are lightweight, especially compared to conventional current transformers with large cores. A higher weight/size advantage arises from the use of combined units, including both current and voltage sensors.

- Safety: Low secondary (transmitted) voltage.

4. Proposed method of protection

Figures 4 and **6** shows a single-phase (internal and external) fault differential protection system. Obviously, a pair of current sensors by using RC surrounded the protected area. Since this was a natural trend, differential protection provided protection for system equipment wire carrying current from the RC called a pilot wire.

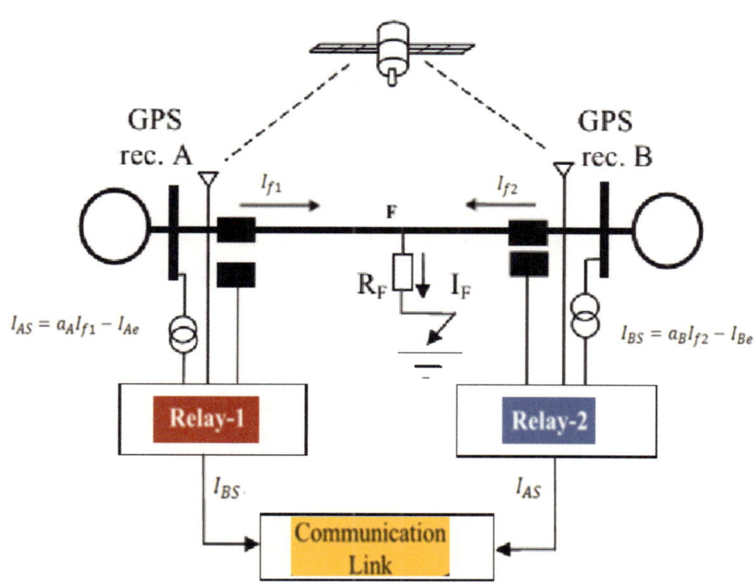

Figure 4.
Differential relay currents at the time internal fault.

In case of no fault, current input protection unit I_p is to be at all times the same to the current going out of the protected zone. Considering current in transformer A, the pilot wire was carrying current as follows:

$$I_{AS} = a_A I_p - I_{Ae} I_{AS} = a_A I_p - I_{Ae} \tag{4}$$

where a_A is the conversion ratio for RC A and I_{Ae} is the secondary excitation current for RC A. Similarly, for current in transformer B, the equation is as follows:

$$I_{BS} = a_B I_p - I_{Be} \tag{5}$$

where a_B is the conversion ratio for RC B and I_{Be} is the secondary excitation current for RC B.Considering the equal ratio, $a_A = a_B = a$, relay working current I_{op} is:

$$I_{op} = I_{Ae} - I_{Be} \tag{6}$$

When there was a normal operation or the external fault of the system, the relay's working current I_{op} was very small, but never tends to zero. Nevertheless, within the time the internal fault was in the protected area, the current was no longer equal to the outgoing current. The operating current of the differential relay is nothing more than an increase of the input current as the same as the feed fault.

$$I_{op} = a(I_{f1} + I_{f2}) I_{Ae} - I_{Be} \tag{7}$$

The uniform distribution of the line is as shown in **Figure 5**. Where, co is a shunt capacitance (F/km), go the shunt leakage conductance (S/km), Io the series inductance (H/km) and ro is a series resistance (Ω/km). The distribution of current and voltage along the transmission line is given by the equations of the current and voltage display in the diagram at the ends of the line.

$$-\frac{\partial u}{\partial x} = r_o i + l_o \frac{\partial i}{\partial t} \tag{8}$$

$$-\frac{\partial i}{\partial x} = g_o u + c_o \frac{\partial i}{\partial t} \tag{9}$$

By reducing Eqs. (5) and (6) to their frequency domain,

$$\begin{bmatrix} U_m \\ I_m \end{bmatrix} = \begin{bmatrix} ch(\gamma l_{nm}) & -Z_c sh(\gamma l_{nm}) \\ sh(\gamma l_{nm})/Z_c & -ch(\gamma l_{nm}) \end{bmatrix} \begin{bmatrix} U_n \\ I_n \end{bmatrix} \tag{10}$$

where $sh(\gamma l_{nm})$ and $ch(\gamma l_{nm})$ = hyperbolic function, Zc = characteristic impedance, γ = the propagation constant and both are freqency dependent; l_{nm} = distance from end to end m.

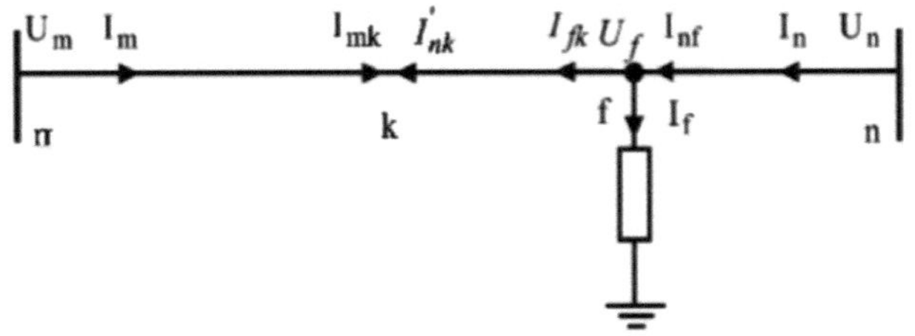

Figure 5.
Transmission line model with a fault located inside the two ends.

By considering the appropriate characteristic impedance and propagation constant, we can apply the above conversion method to the zero, positive and negative sequence components of the current. In this paper, the "0," "1," and "2" subscripts were described as zero, positive, and negative sequences, respectively.

$$I_{mk0} = I_{m0}ch(\gamma_0 l_{mk}) - \frac{U_{m0}}{Z_{c0}}sh(\gamma_0 l_{mk}) \tag{11}$$

$$I_{mk1} = I_{m1}ch(\gamma_1 l_{mk}) - \frac{U_{m1}}{Z_{c1}}sh(\gamma_1 l_{mk}) \tag{12}$$

$$I_{mk2} = I_{m2}ch(\gamma_1 l_{mk}) - \frac{U_{m2}}{Z_{c2}}sh(\gamma_1 l_{mk}) \tag{13}$$

$$I_{nk0} = I_{n0}ch(\gamma_0 l_{nk}) - \frac{U_{n0}}{Z_{c0}}sh(\gamma_0 l_{nk}) \tag{14}$$

$$I_{nk1} = I_{n1}ch(\gamma_1 l_{nk}) - \frac{U_{n1}}{Z_{c1}}sh(\gamma_1 l_{nk}) \tag{15}$$

$$I_{nk2} = I_{n2}ch(\gamma_1 l_{nk}) - \frac{U_{n2}}{Z_{c2}}sh(\gamma_1 l_{nk}) \tag{16}$$

where l_{mk} and l_{nk} = distance of point k from end m and n respectively.

$$a = e^{j2\pi/3}$$

$$\begin{bmatrix} I_{mka} \\ I_{mkb} \\ I_{mkc} \end{bmatrix} = \begin{bmatrix} 1 & 1 & 1 \\ 1 & a^2 & a \\ 1 & a & a^2 \end{bmatrix} \begin{bmatrix} I_{mk0} \\ I_{mk1} \\ I_{mk2} \end{bmatrix} \tag{17}$$

Similarly, the relationship to current is as below:

$$\begin{bmatrix} I_{nka} \\ I_{nkb} \\ I_{nkc} \end{bmatrix} = \begin{bmatrix} 1 & 1 & 1 \\ 1 & a^2 & a \\ 1 & a & a^2 \end{bmatrix} \begin{bmatrix} I_{nk0} \\ I_{nk1} \\ I_{nk2} \end{bmatrix} \tag{18}$$

If the fault is outside the protected area or there is no fault in the system, then the phase current should satisfy the following equation:

$$I_{mk\varnothing} = -I_{nk\varnothing} \tag{19}$$

where $\varnothing$ = a, b, c shows the phase from where the current belong to.

The following explanation makes it possible to understand the actual current and the position where the fault and the derivation of the current occur. To calculate the fault location of the current and voltage from the end n,

$$U_{f0} = U_{n0}ch(\gamma_0 l_{nf}) - I_{n0}Z_{c0}sh(\gamma_0 l_{nf}) \tag{20}$$

$$U_{f1} = U_{n1}ch(\gamma_1 l_{nf}) - I_{n1}Z_{c1}sh(\gamma_1 l_{nf}) \tag{21}$$

$$U_{f2} = U_{n2}ch(\gamma_1 l_{nf}) - I_{n2}Z_{c1}sh(\gamma_1 l_{nf}) \tag{22}$$

$$I_{f0} = I_{n0}ch(\gamma_0 l_{nf}) - \frac{U_{n0}}{Z_{c0}}sh(\gamma_0 l_{nf}) \tag{23}$$

$$I_{f1} = I_{n1}ch(\gamma_1 l_{nf}) - \frac{U_{n1}}{Z_{c1}}sh(\gamma_1 l_{nf}) \tag{24}$$

$$I_{f2} = I_{n2}ch(\gamma_1 l_{nf}) - \frac{U_{n2}}{Z_{c2}} sh(\gamma_1 l_{nf}) \tag{25}$$

The differential current is the actual measure of relay operation. In addition, the brake current is one of the avoided currents of the mal-tripping differential relay. The effect of this current has been in breaking the relay.

$$I_B = |I_{mk1} + I_{nk1}| \tag{26}$$

$$I_B = \left| I_{mk1} + I_{nk} - I_{f1} ch(\gamma_1 l_{fk}) \right| \tag{27}$$

The effects of the capacitive current can be cancelling with any of the use of the following techniques. Shunt Reactor, Phasor Compensation algorithm, Capacitor Current Compensation, and Bergeron line Model.

In reality, the proposal is a reliable method for the protection system. Since each current transformer and circuit breaker can only declare a line, it is usual to protect by the use of two side current transformers and circuit breakers. The currents on both sides were compared. Under normal conditions or for defects outside the protected area, the current Bas-Bar A is equal to the current Bas-Bar B. Thus, the currents in the secondary current transformer were equal; no current flowed through the relay current. If a fault occurs in the protected zone, the secondary transformer currents of Bas-Bar A and Bas-Bar B will not be the same, and there will be a current flowing through the relay current, as shown **Figure 6**.

The differential protection ratio, by using a multi-slope feature of the relay was inserted into the new relay for its excellent compromise between reliability and sensitivity. The components of the relay compared the different current I_{diff} (also called operating current), and the restraining current I_{bias} are expressed in Eqs. (28) and (29) [20–24].

$$I_{diff} = |I'_{As} + I'_{Bs}| \tag{28}$$

$$I_{bias} = (|I'_{As}| + |I'_{Bs}|)/2 \tag{29}$$

where I'_{As} and I'_{Bs} are secondary RC phasor currents.

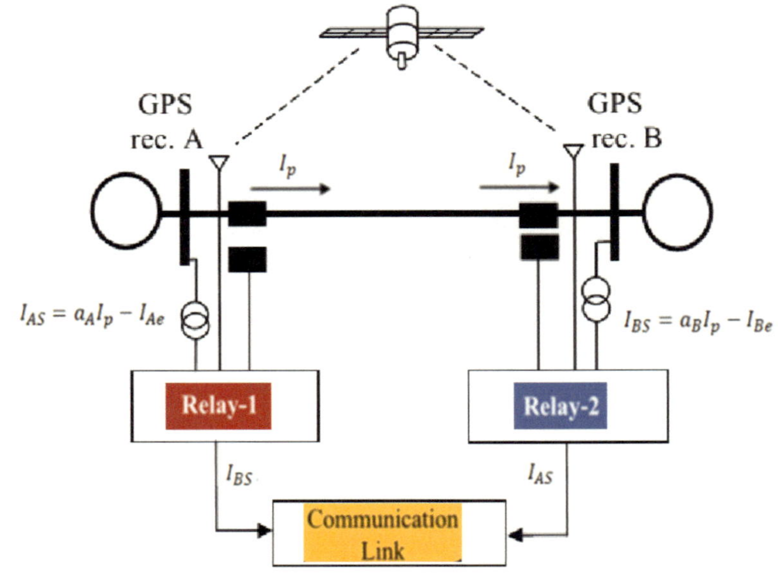

Figure 6.
Differential relay currents at the external fault.

5. SCADA systems

It is considered as a means of monitoring and control of power plants also used in renewable power plants and these systems transfer data to the heart of the system which is a master computer and receives orders from many remote terminals; **Figure** 7 illustrates the structure of the SCADA system, the SCADA system includes the following:

1. RTUs or PLC which transfers information to the central unit and transfers orders to equipment.

2. Radio system or satellites to secure communication between central units and distributed areas away from the focal point.

3. The software package used in the system.

The operation of electricity distribution networks is monitored and controlled by supervisory control and data acquisition (SCADA) systems. SCADA systems are linked through various communications networks such as microwave and optical fiber networks to ensure the functioning of the system. They are used to connect transmission substations with the major generators to facilitate an integrated system. The operation of the methods of communication in the networks of energy is produced by lines along the system with advanced optical networks. The loss of these communication cables is possible and could make the protection and monitoring of the network more complex. Using advanced wireless communication and sensing devices could improve the control and monitoring of the entire system. Intelligent Electronic Devices (IED) for monitoring and control to improve the technology of smart grids, and. IEDs can be installed and distributed within the system to be used for protection and monitoring [24–28]. These devices are

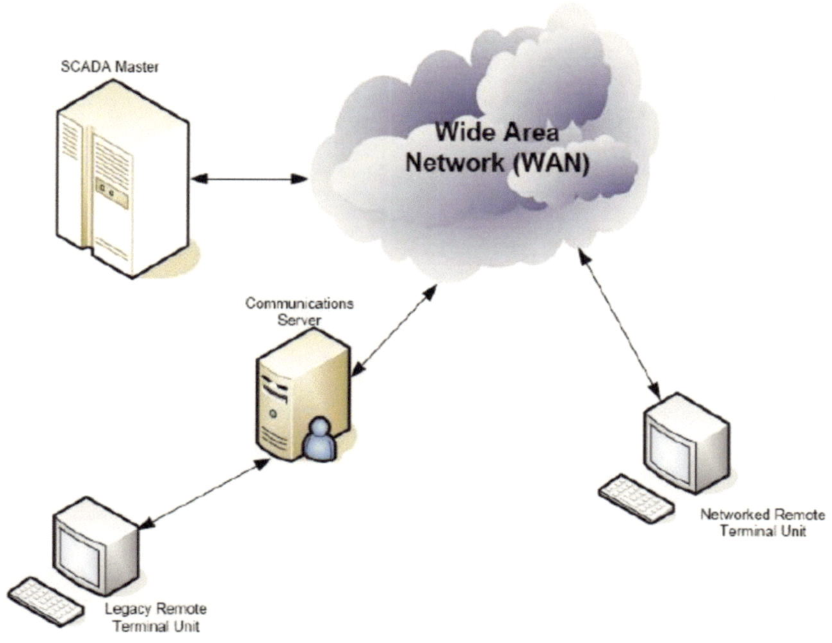

Figure 7.
Network monitoring and control by supervisory control and data acquisition systems (SCADA).

interconnected and can be communicated to the central IED, which is implemented in the substation. On the other hand, IEDs can also monitor and update the electric flow of real-time status and can be used to manage and control the network. **Figure 7** shows the monitoring and control by supervisory control and data acquisition system for smart grid technology.

6. System modeling

The simulation of line differential protection is presented in Simulink/MatLab environment, as shown in **Figure 8**.

It simulates three-phase, the system components are a transmission line with PV arrays (800 kW), Wind farm (9 MW), Resistive load (45 kW, 100 kW), 25 kV distribution Bus, and the utility grid. The proposed settings of the protection scheme for transmission lines from Bus-Bar 7 to Bus-Bar 9, and the information corresponding to the lines are listed in **Table 2**, where is the rated secondary current. In this study, is taken to be 1 Amp. According to the value of, the following constants are:

$$I_O = 0.3, \quad I_{S2} = 0.2, \quad K_1 = 0.35, \quad K_2 = 1.2;$$

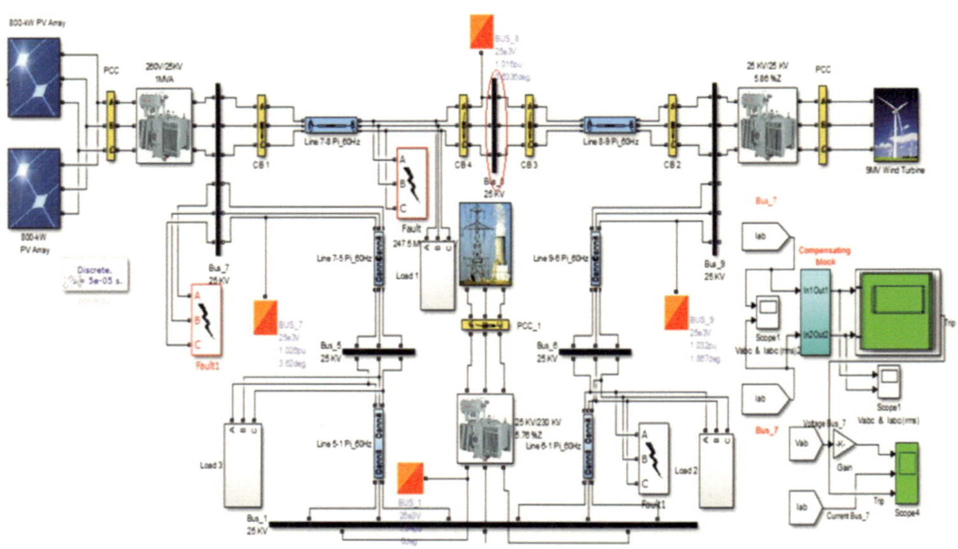

Figure 8.
Simulation representation of the differential protection scheme.

Relay setting	Range
Differential current setting	(0.2–2.0 *In*)
Bias current threshold setting	(1–30)
Lower percentage bias setting	(0.3–1.5)
Higher percentage bias setting	(0.2–2.0 *In*)

Table 2.
Relay setting ranges.

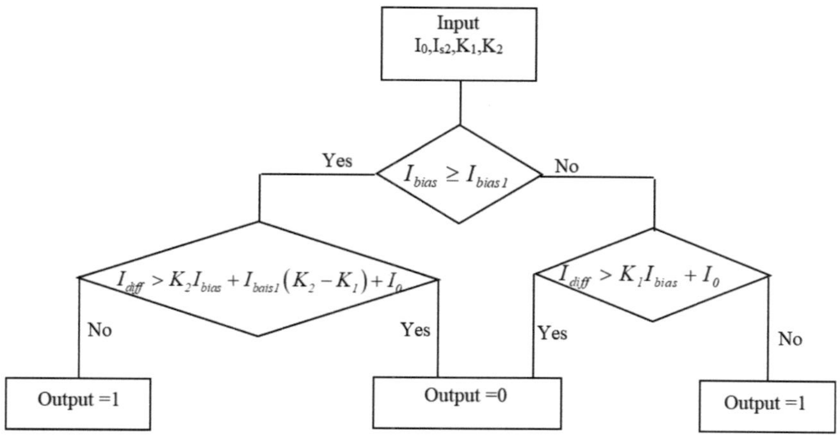

Figure 9.
Flowchart for decision block.

where *In* is the current rating of the secondary current transformer; I_o is the differential current at zero bias current; *Ibias* is the bias current when the relay characteristic starts to change; and K_1, K_2 is the percentage biases. These constants can be obtained from the relay characteristic, which is used in relay operations. The differential protection operating region is above the slope of the characteristic, and the restraining of the region is below the slope of the characteristic. The dual slope bias technique is used to improve stability through fault and external fault *CT* saturation to provide further security. To achieve the appropriate setting, the characteristics of the relay to be applied must be considered. Three adjustable values (from the relay manual) are recommended as follows.

The block consists of a relay that is divided into two units, as shown in **Figure 9**.

7. Simulation results and discussion

Differential protection between the two circuit breaker CBs, are known the local side (CB1) and far side (CB2), the Current Sensors RCA and RCB are installed at the Bus-Bar (7 and 9), to measure the current and voltage per phase. The current and voltage of each phase of the analog signal are converted into digital data using an analog-to-digital converter (ADC). The local side is directly connected to the relay, whereas the far side is connected through the delay block. Channel time delay is set at 27 ms, the output of the differential relay block is binary (0, 1). When disturbances happen in the protected zone, a current difference between the two CBs (local and remote) exists. Thus, the relay sends a signal tripping to isolate the defective zone from the rest of the protection scheme. In this study, we propose that the fault will occur in the middle of the transmission line, starting at 0.3 ms. In the fault parameter block, the fault type is simulated separately using MATLAB. This section investigates the cases of faults (internal and external).

Five cases have been tracked: Case 1, single line to ground fault (SLGF); Case 2, line-to-line fault (LLF); Case 3, double line to ground fault (2LGF); Case 4, three-line fault (3LF); and Case 5, three line to ground fault (3LGF). The main principle of the differential relay should be compared for both ends of the protected area. Modeling for each case is similar through comparison of the current at both ends.

All used case results are indicated in **Figures 10–24**. Three diagrams, namely, remote, local, and differential, are presented in every figure. Remote and local measurements are conducted separately; each colour in this diagram represents one

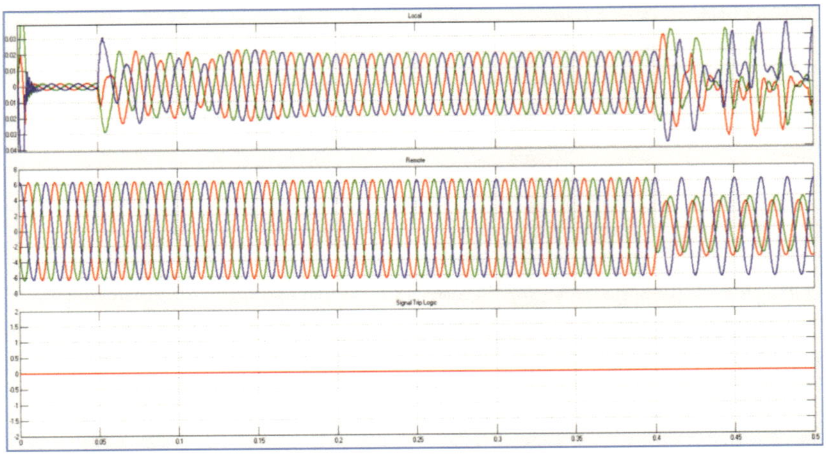

Figure 10.
SLGF current waveform.

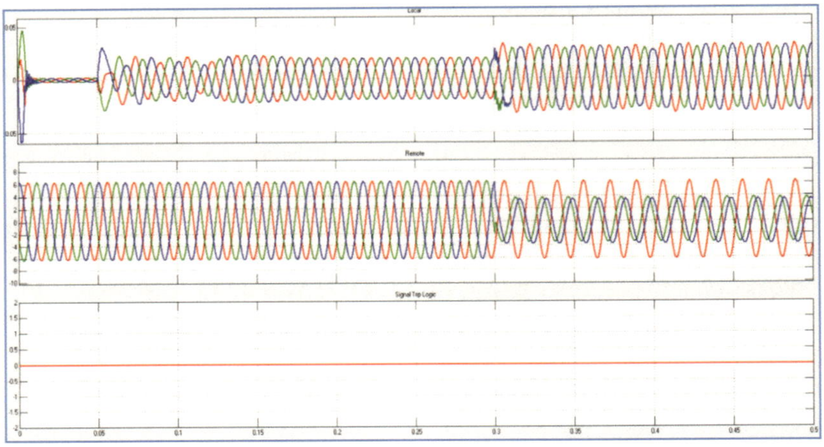

Figure 11.
SLGF voltage waveform.

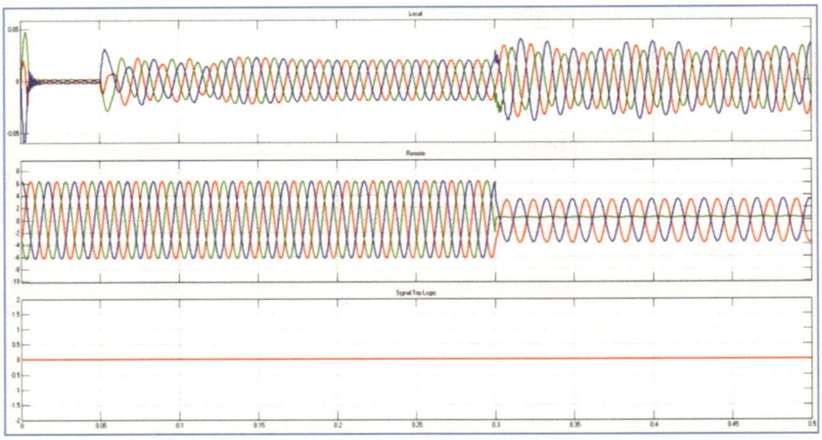

Figure 12.
2LGF current waveform.

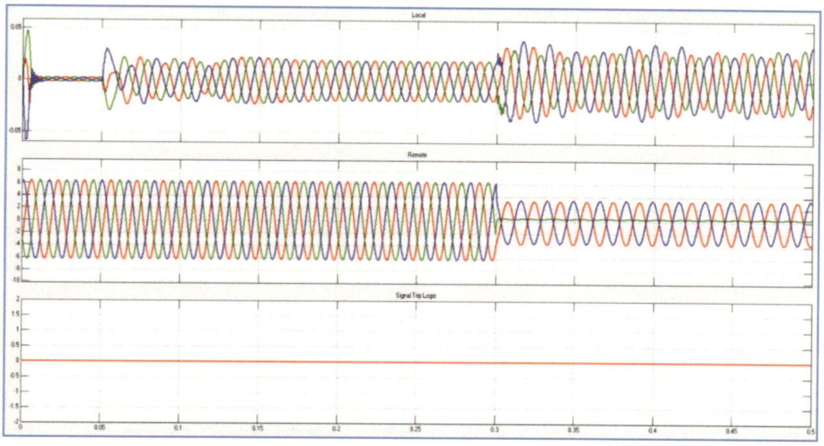

Figure 13.
2LGF voltage waveform.

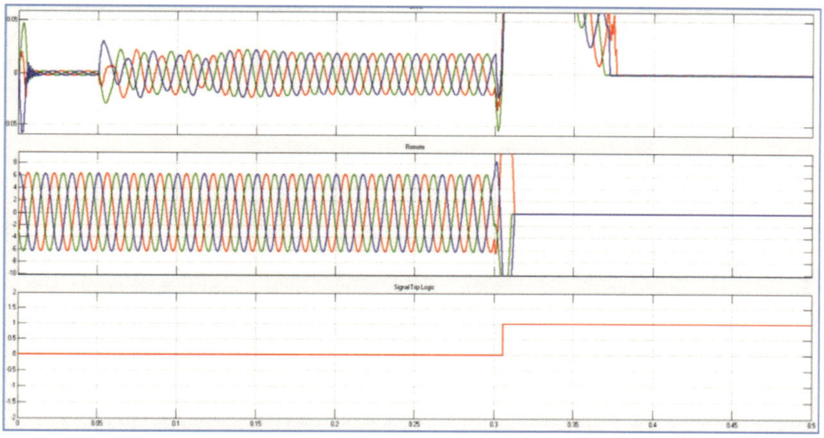

Figure 14.
SLGF (A) current waveform.

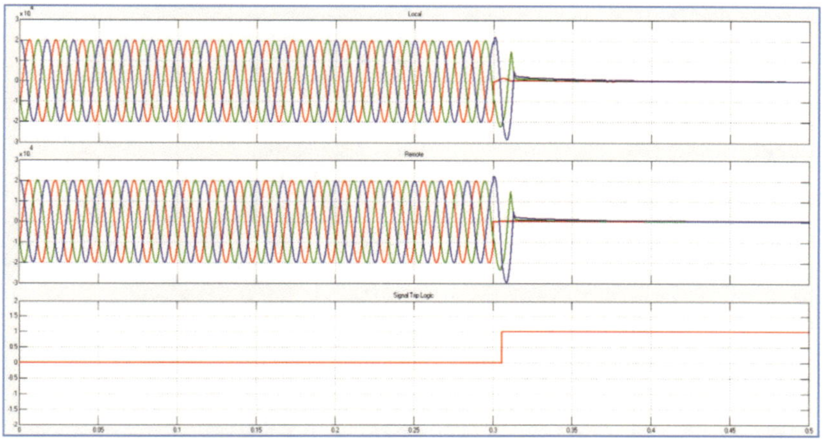

Figure 15.
SLGF (A) voltage waveform.

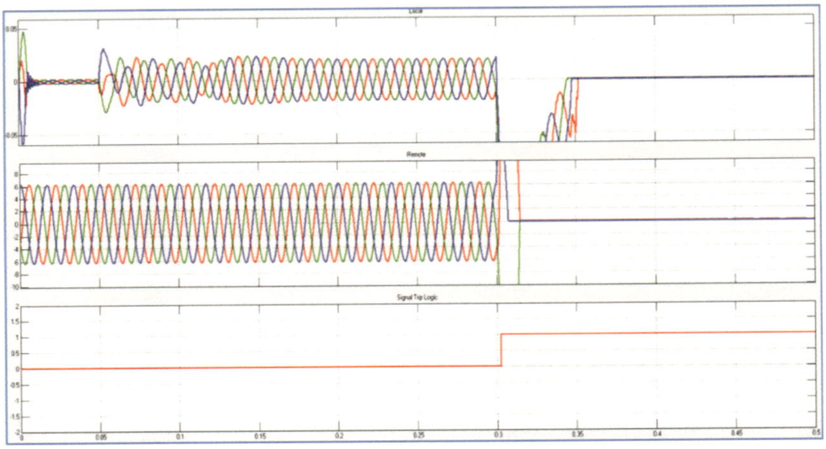

Figure 16.
Double-line (A and B) fault current waveform.

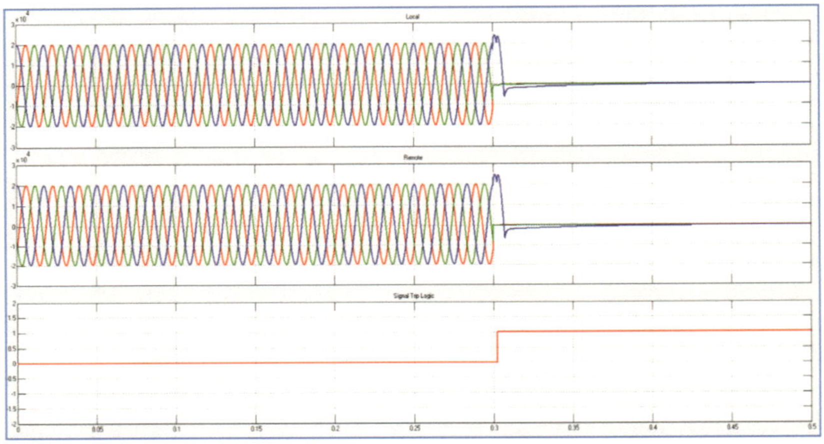

Figure 17.
Double-line (A and B) fault voltage waveform.

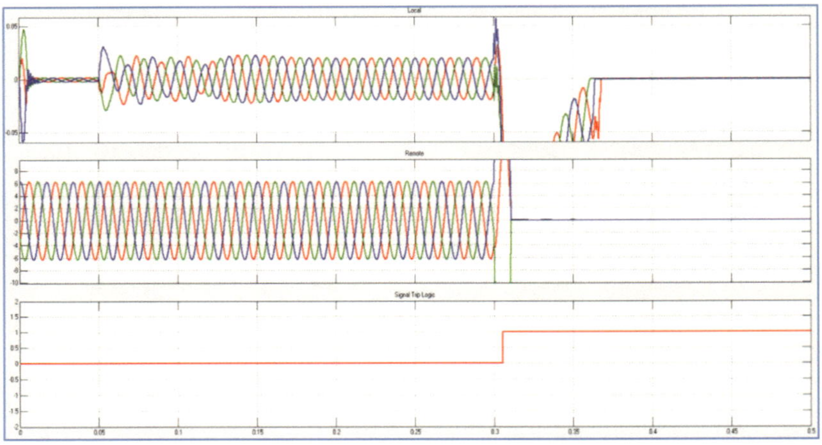

Figure 18.
2LGF (B and C) current waveform.

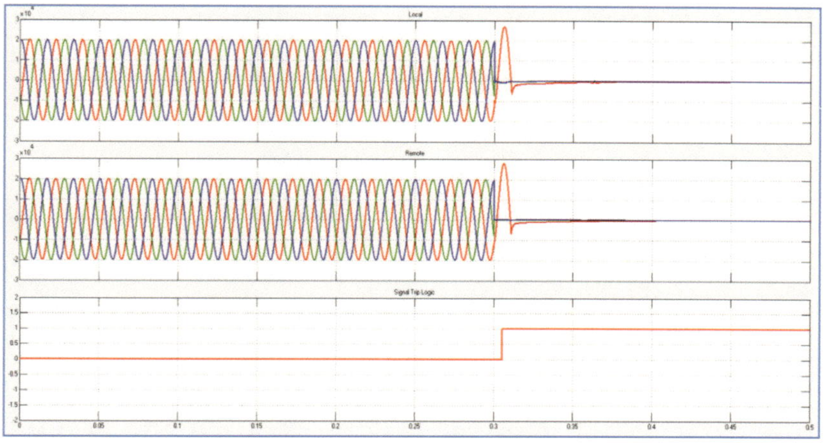

Figure 19.
2LGF (B and C) voltage waveform.

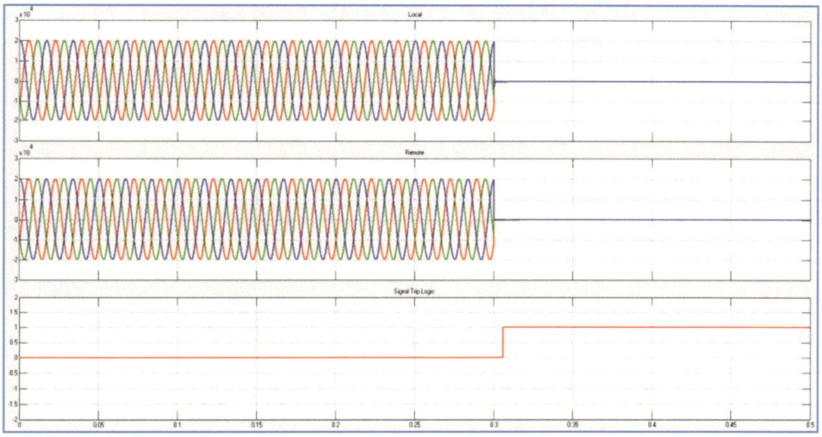

Figure 20.
3LF current waveform.

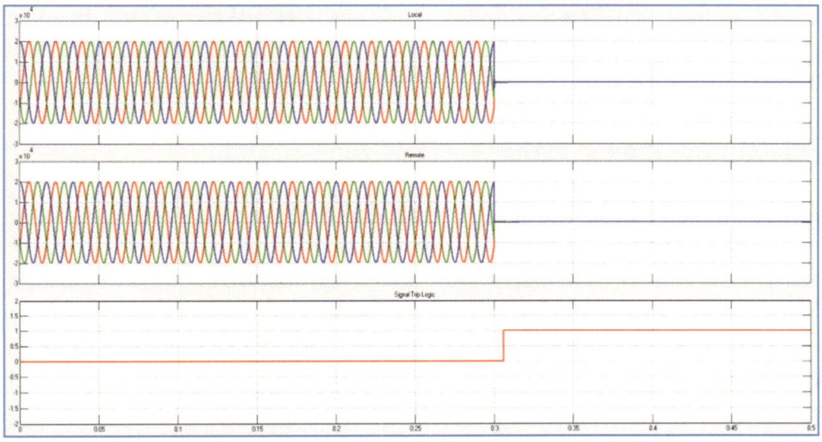

Figure 21.
3LF voltage waveform.

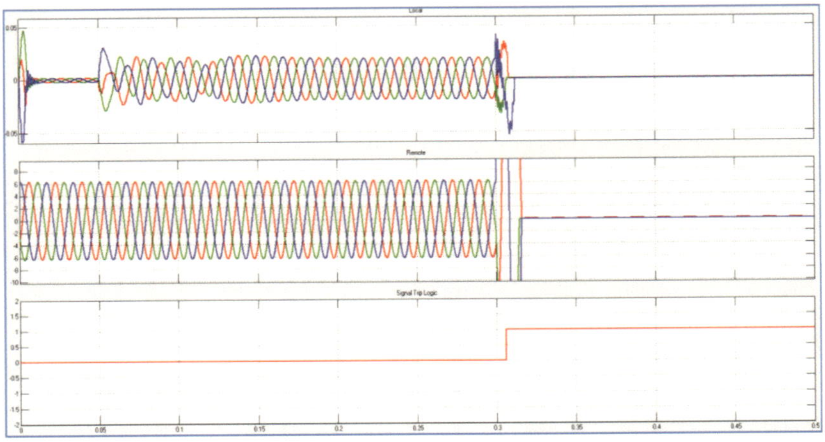

Figure 22.
3LGF current waveform.

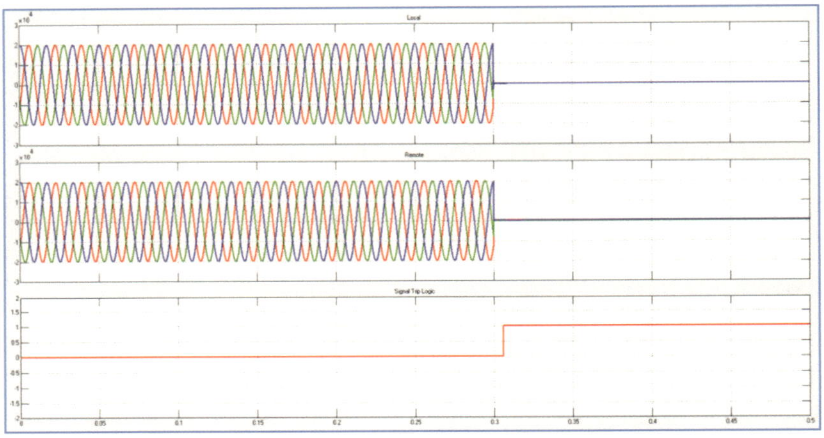

Figure 23.
3LGF voltage waveform.

phase, namely, red, green, and blue for phases A, B, and C, respectively. The signal to Trip is represented as a binary (0, 1) measurement depending on the difference between the signals of the two ends for sending the signal to the circuit breaker. In **Figures 10–23**, two types of status are shown: for current differential and the voltage differential, with the fault occurring at t = 0.3 ms.

7.1 Simulation result outside the protected zone

Figures 10 and **11** show the results of an external L-G fault on protection scheme. That is differential relays without sending tripping signal.

Similarly, the results of external fault on the protection scheme in various fault conditions, that is, that mean the relay No sending trip signal to the circuit breaker.

7.2 Simulation result within the protected zone

Figures 14–23 show the results of an internal fault for the protected zone. The relay sends a trip signal to the circuit breaker under a faulty condition, and the circuit breaker isolates the zone relay from the rest of the protection scheme.

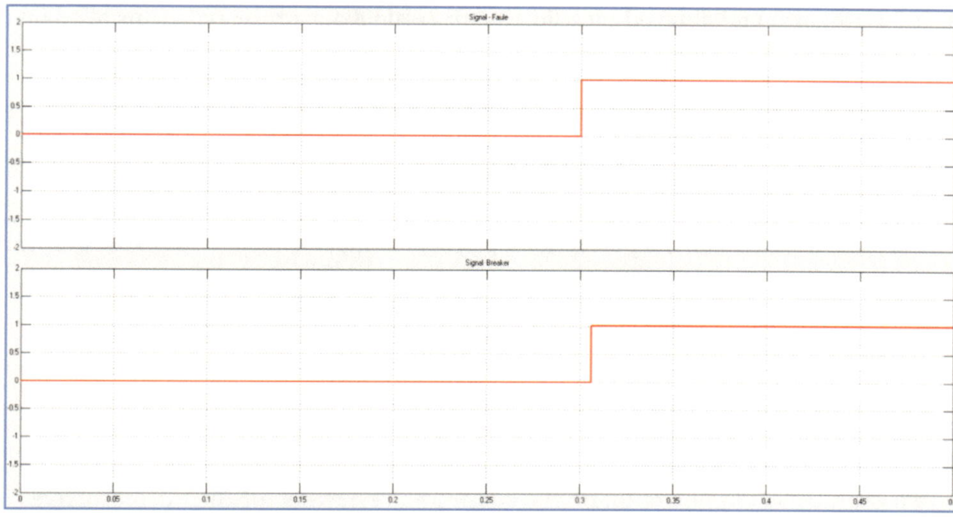

Figure 24.
The difference between the occurrence of a fault and a breaker signal.

Measurements	Zone relay
Relay output	Trip or no trip
The threshold of operation	t = 0.3 ms
Decision speed	Less than one cycle

Table 3.
Specifications of the proposed relay.

Figure 24 shows the time differences between the moment of a fault and signal of the breaker during and after the occurrence of the fault (**Table 3**). Displays the output of our proposed relay measurements; compared with the traditional behaviour in normal operation, the speed of our proposed scheme is less than one cycle.

8. Summary

The reliability of the relay protection system can be described in two respects: dependability and security. The reliability of the relay protection system detects and disconnects all faults in the protection zone. The safety of the relay protection system is capable of rejecting all events and transients that are not faulty so that the healthy part of the power system is not unnecessarily disconnected. Differential protection is the preferred solution for widespread use; fault protection for multi-terminal systems becomes very difficult, and fast fault detection of systems becomes very important. This result provides different solutions for transmission line protection. This method is better than distance protection because differential protection requires fewer input data and reduces computation time.

The performance of this algorithm is more efficient than distance relay protection. The disadvantages of the distance on the transmission line and the directional over-current relay are as follows:

1. If a fault occurs at the end of the line, the relay cannot be disconnected immediately at both ends of the line.

2. Coordination is achieved by adjusting the time delay of the relays installed on the power line beside the primary protection and backup protection. Therefore, the delay time of the relay operating in each protection zone will slow down the termination of the interference.

The differential protection principle is based on Kirchhoff current law, which has been widely used in the primary equipment protection of the power systems. The general objective of the protection system is to quickly isolate the areas that contain unrest while preserving the rest of the system. The method of protection must meet five criteria to perform successfully: (1) reliability, (2) selectivity, (3) speed, (4) simplicity, and (5) economy.

Acknowledgements

The author would like to thank the editor and reviewers for their constructive comments and suggestions on this article. This work was supported by the Southern Technical University-Iraq, BETC, and the Huazhong University of Science and Technology-China (no. 01122018).

Conflicts of interest

The authors declare no conflicts of interest.

Author contributions

I would like to express my deep appreciation to Ms. Janan Abd Ali al-Hajji, also to Dr. Shaorong Wang (HUST), who oversaw my studies, and to the editor of books IntechOpen Dr. Ahmed Zobaa and Dr. Alfredo Vaccaro, I have the honor of working with them and join them.

Author details

Ali Hadi Abdulwahid[1,2]* and Adnan A. Ateeq[2]

1 School of Electrical and Electronic Engineering, Huazhong University of Science and Technology, Wuhan, Hubei, China

2 Department of Engineering Electrical Power, Engineering Technical College, Southern Technical University, Basra, Iraq

*Address all correspondence to: dr.alhajji_ali@yahoo.com

References

[1] Ahmad A, Hassan NU. Smart Grid as a Solution for Renewable and Efficient Energy. 2016

[2] Tong C, Wang Q, Gao Y, Tong M, Luo J. Dynamic lightning protection of smart grid distribution system. Electric Power Systems Research. 2014;**113**: 228-236

[3] Muhammad Ramadan BMS, Chiang Jia Hao E, Logenthiran T, Naayagi RT, Woo WL. Islanding detection of distributed generator in presence of fault events. In: Proceedings of the IEEE Region 10 Conference (TENCON); 5-8 November 2017

[4] Muhammad Ramadan BMS, Surian R, Logenthiran T, Naayagi RT, Woo WL. Self-healing Network Instigated by Distributed Energy

[5] Muhammad Ramadan BMS, Logenthiran T, Naayagi RT, Su C. Accelerated lambda iteration method for solving economic dispatch with transmission line losses management. In: Proceedings of IEEE PES Innovative Smart Grid Technologies (ISGT Asia) Conference; 2016. pp. 138-143

[6] Logenthiran T, Naayagi RT, Woo WL, Phan V-T, Abidi K. Intelligent control system for microgrids using multi-agent system. IEEE Journal of Emerging and Selected Topics in Power Electronics. 2015;**3**(4):1036-1045

[7] Zhang J, Dong Y. Power-aware communication management for reliable remote relay protection in smart grid. In: IEEE Power Systems Conference; 2016. pp. 1-6

[8] Artale G, Cataliotti A, Cosentino V, Di D, Nguyen N, Russotto P, et al. Hybrid passive and communications-based methods for islanding detection in medium and low voltage smart grids. In: Proceedings of International Conference on Power Engineering Energy and Electrical Drives Powereng-2013; May 2013. pp. 1563-1567

[9] Artale G, Cataliotti A, Cosentino V, Di D, Nguyen N, Tinè G. Measurement and communication interfaces for distributed generation in smart grids. In: Proceedings of 2013 IEEE International Workshop on Applied Measurements for Power Systems AMPS; 2013. 2013. pp. 103-107

[10] Safdar S, Hamdaoui B, Cotilla-Sanchez E, Guizani M. A survey on communication infrastructure for micro-grids. In: Wireless Communications and Mobile Computing Conference (IWCMC) 2013 9th International; July 2013. pp. 1-5

[11] Pengcheng Z. IEC 61850-9-2 process bus communication Interface for light weight merging unit testing environment: Thesis. Stockholm, Sweden: KTH Electrical Engineering; 2012

[12] Wibowo H, Ary PN, Hidayat S. Design and testing of Rogowski coil based PCB double helix for gas insulated switchgear 150 kV aplication. In: Power Engineering and Renewable Energy (ICPERE); 2016 3rd Conference on. IEEE. 2017

[13] Elkalashy NI, Kawady TA, Esmail EM, Taalab AI. Fundamental current phasor tracking using DFT extraction of Rogowski coil signal. IET Science, Measurement and Technology. 2016; **10**(4):296-305

[14] Hajipour E, Vakilian M, Pasand MS. Current-transformer saturation compensation for transformer differential relays. IEEE Transactions on Power Delivery. 2015;**30**(5):2293-2302

[15] Metwally IA. Novel designs of wideband Rogowski coils for high

pulsed current measurement. IET Science, Measurement and Technology. 2014;**8**(1):9-16

[16] Hemmati E, Shahrtash SM. Digital compensation of Rogowski coil's output voltage. IEEE Transactions on Instrumentation and Measurement. 2013;**62**(1):71-82

[17] Al-Sowayan S. Improved mutual inductance of Rogowski coil using hexagonal core. International Journal of Electrical and Computer Engineering Electronic and Communication Engineering. 2014;**8**(2):293-296

[18] Marracci M, Tellini B. Analysis of precision Rogowski coil via analytical method and effective cross section parameter. In: Proc. IEEE I2MTC; 23–26 May 2016; 2016. pp. 1-5

[19] Vourc'h E, Yu W, Joubert P-Y, Revol B, Couderette A, Cima L. Neel effect toroidal current sensor. IEEE Transactions on Magnetics. 2013;**49**(1): 81-84

[20] Raspolli AM, Antonetti C, Marracci M, Piccinelli F, Tellini B. Novel microwave-synthesis of Cu nanoparticles in the absence of any stabilizing agent and their antibacterial and antistatic applications. Applied Surface Science. 2013;**280**:610-618

[21] Ramesh K, Sushama M. Power transformer protection using fuzzy logic based-relaying. In: International Conference on Advances in Electrical Engineering; 2014. pp. 1-7

[22] Deshmukh MS, Barhate VT. Transformer protection by distinguishing inrush and fault current with harmonic analysis using fuzzy logic. In: IEEE International Conference on Control and Robotics Engineering; April 2016

[23] Barhate V, Thakre KL, Deshmukh M. Adaptable differential relay using fuzzy logic code in digital signal controller for transformer. In: 57th International Scientific Conference on Power and Electrical Engineering of Riga Technical University (RTUCON); October 2016

[24] Paz MCR, Leborgne RC, Bretas AS. Adaptive ground distance protection for UPFC compensated transmission lines: A formulation considering the fault resistance effect. International Journal of Electrical Power & Energy Systems. 2015;**73**:124-131

[25] Han KL, Cai ZX, Xu M, Hi Z. He dynamic characteristics of characteristic parameters of traveling wave protection for HVDC transmission line and their setting. Power System technology. 2013; **37**:255-260

[26] Tavares KA, Silva KM. Modeling and simulation of the differential protection of power transformers in EMTP Softwares. In: DPSP: International Conference on Developments in Power System Protection; 2012

[27] Alvarenga MTS, Vianna PL, Silva KM. High-impedance bus differential protection modeling in ATP/MODELS. Engineering. 2013;**5**:37-42

[28] Franklin R, Nabi-Bidhendi H, Thompson MJ, Altuve HJ. High-impedance differential applications with mismatched CTs. In: Proceedings of the 44th Annual Western Protective Relay Conference; October 2017